The Institute of Biology's
Studies in Biology No. 50

Pest Control
and its Ecology

Helmut F. van Emden

Ph.D.
Reader in Departments of Agriculture & Horticulture and Zoology
University of Reading

Edward Arnold

First published 1974
by Edward Arnold (Publishers) Ltd,
25 Hill Street, London W1X 8LL

Boards edition ISBN: 0 7131 2471 7
Paper edition   ISBN: 0 7131 2472 5

Reprinted 1976

Printed in Great Britain by
The Camelot Press Ltd, Southampton

General Preface to the Series

It is no longer possible for one textbook to cover the whole field of Biology and to remain sufficiently up-to-date. At the same time teachers and students at school, college or university need to keep abreast of recent trends and know where the most significant developments are taking place.

To meet the need for this progressive approach the Institute of Biology has for some years sponsored this series of booklets dealing with subjects specially selected by a panel of editors. The enthusiastic acceptance of the series by teachers and students at school, college and university shows the usefulness of the books in providing a clear and up-to-date coverage of topics, particularly in areas of research and changing views.

Among features of the series are the attention given to methods, the inclusion of a selected list of books for further reading and, wherever possible, suggestions for practical work.

Reader's comments will be welcomed by the author or the Education Officer of the Institute.

1974   The Institute of Biology,
                                              41 Queens Gate,
                                              London, SW7 5HU

Preface

This slim book cannot attempt to cover pest control in its full diversity; it is limited to the clearly definable area of the control of insect pests on growing crops. There is more to such control today than applying poisonous chemicals. The questions that have to be asked are largely biological and ecological, and their answer affects our food supply, our standard of living and our environment. The text suggests many studies that can be attempted around any farm or in a simple laboratory or glasshouse.

This is not really a book about ecology; it is firmly about pest control. It tries to assess where we are going, and the ecological merits, record to date and future potential of the many approaches. Such an assessment has strong subjective overtones, and another author might well make a contrasting judgement.

Pest control is indeed a 'Study in Biology'. Yet the biologist, and particularly the ecologist, who is going to play an increasing role in the subject in the future, will have maximum impact if he can visualize the total environment of crops to include not only soil, weather, insects and plants but also the social, economic and industrial environments with which the crops are inexorably enmeshed.

Reading, 1974   H. v. E.

Contents

1 Brief Introduction to Chemical Control

The history of modern pest control dates from the second half of the nineteenth century, when Colorado beetle *(Leptinotarsa decemlineata)* (see section 2.5) spread rapidly across the United States and seriously threatened food production and the national economy. After much argument it was finally decided to take the unprecedented step of spraying the potato crops with a human poison (arsenic in the form of Paris Green). It would be inhuman to claim it as unfortunate that the prophesied human mass mortality did not occur, but there is no doubt that Colorado beetle control with Paris Green opened the way to the use of biocides (destroyers of life in general) on crops destined for human consumption.

1.1 Insecticides

1.1.1 The early insecticides

These were mainly of three kinds. One group were the stomach poisons (including Paris Green), toxic radicals (arsenic or fluorsilicates) formulated as salts of metals (particularly lead and sodium). Such salts were relatively stable and plants could be sprayed with the compounds without symptoms of phytotoxicity appearing. Parasites and predators could contact the pesticide deposit without being killed. However, when a leaf-feeding pest consumed treated parts of the plant, the pesticide was hydrolysed in the gut to release free toxin (e.g. arsenic) which precipitated the insect protoplasm and the pest was thereby killed.

The second group were tar oils which, owing to their deleterious effects on plants, could only be used on dormant plants without leaves such as fruit trees in winter. They were effective in coating and killing the eggs and overwintering larvae of many insects.

The third group were toxic extracts of plants as long used by primitive tribes to tip their hunting arrows or bring fish to the surface of rivers and lakes. Best known of these substances are pyrethrum (from a chrysanthemum-like plant), rotenone (a root extract of the derris plant) and nicotine (from tobacco). These extracts work in various ways, either paralysing the nervous system or checking the insect's breathing. An important feature of these insecticides is that they penetrate the cuticle and therefore do not need to be ingested like the stomach poisons; also they are very short-lived (hours or days) and therefore the insect has to be contacted by drops of the spray (ephemeral contact insecticides). At present, there is a new and considerable interest in synthesizing pyrethroid insecticides. These appear capable of being given many desirable characteristics, particularly through additives. Although they have not as yet attained wide commercial use, they may prove to be the insecticides of the future.

1.1.2   Residual contact insecticides (Fig. 1–1)

Farmers and growers were naturally very interested when longer lasting synthetic poisons became available and a crop could be protected against insects for up to several weeks or months. Moreover, it then became less important to contact the insect at the time of application; it could be killed later when, in walking, it contacted the residual film of pesticide laid down on the plant surface.

The first residual contact insecticides were the organochlorines. Many of these, e.g. DDT, BHC, Heptachlor, Aldrin and Dieldrin, were used very extensively as emulsions in water, wettable powders, harrowed into the soil or coated on seeds (seed dressings). They are not unduly toxic to man, but many are extremely persistent and some probably remain in the environment for periods approaching the life-span of man. They give long lasting and inexpensive control of many pests other than the small sessile sucking insects such as aphids and scale insects. The mode of action of this group of poisons is as yet not fully understood, but major effects appear to be an inhibition of the enzyme cytochrome oxidase, which mediates gas exchange in the respiration of all animals which use blood circulation as a gas carrier, and a poorly understood de-stabilization of the nervous system. The first organochlorine insecticides were produced in the late 1930s, were the predominant insecticides in use in the 1940s and 1950s, and many are still in use today.

A second group of residual contact insecticides was produced in the early 1950s. These are the organophosphorus compounds including Parathion, Malathion, TEPP, Dimeton, Metasystox, Schradan and Diazitol. They differ from the organochlorines in being much less persistent (lasting only a few days), and are generally much more poisonous to both insects and man. Like the organochlorines, the organophosphates have a very non-specific mode of action on animals. They combine with the enzyme acetylcholinesterase, and thus inhibit the hydrolysis of the acetylcholine produced at the nerve endings to carry nerve impulses across the synapses. In poisoned animals, therefore, acetylcholine accumulates in the synapses, giving constant nervous stimulation resulting in tetanic paralysis. Several organophosphates are absorbed through the leaves and roots of plants and then translocated. Such insecticides are said to be 'systemic'. The plant compensates for poor pesticide coverage by re-distributing the chemical, and organs produced after spraying may be protected; the systemic action is especially useful against aphids and other insects which take up plant sap. This sap will be poisonous, yet the plant surface may be quite safe for other insects, including parasites and predators. The organophosphates are the principal pesticides in use today, having gradually gained popularity over the persistent organochlorines in the course of the 1960s. It has been possible to replace organochlorines with organophosphates even for some soil applications where long persistence is required. The organophosphate (usually Diazitol) is incorporated in a granule which can be placed on or in the soil. The relatively non-persistent chemical is continually replaced in the soil as it is slowly leached from the granule by the soil water. Organophosphates can also be made to mimic stomach poisons (i.e. toxic on ingestion and therefore selective for plant-feeding insects) by encapsulating minute droplets in a shell of inert material. These micro-capsules can be sprayed, but the technique is still on the research bench.

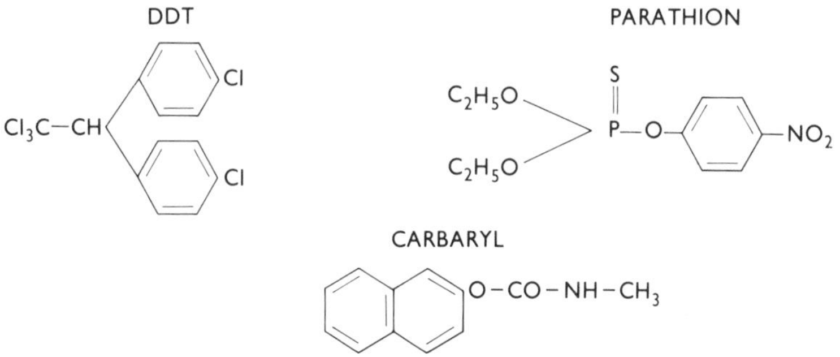

Fig. 1–1   Chemical structure of the three main groups of residual contact insecticides: Organochlorine (e.g. DDT); Organophosphorus (e.g. Parathion); Carbamate (e.g. Carbaryl).

A third group of residual contact insecticides, derived from the motor tyre industry, appeared in the 1960s. These are the carbamates, whose persistence and toxicity lie intermediate between the organochlorines and the organophosphates. Carbaryl (Sevin) is perhaps the best known; it has been widely used for the control of caterpillars and other surface feeders, but recently a wider potential for the carbamates has been realized with the development of a systemic carbamate with excellent aphicidal properties. The action of the carbamates, like that of the organophosphates, is on the nervous system by the accumulation of acetylcholine at the nerve synapses. Rather than inhibiting acetylcholinesterase, they perhaps act as competitors with acetylcholine for the enzyme's surface.

1.2   The application of pesticides

In brief, pesticides may be applied in one of three ways:

(a) spraying liquid droplets with the compound dissolved or as a small particle (wettable powder), with water or air as the diluent;

(b) spreading the compound absorbed on to an inert solid carrier which acts as the diluent (dusts);

(c) burning the compound to create a pesticidal smoke which will penetrate all parts of a more or less enclosed space (e.g. a dense orchard, a glasshouse).

The subject of pesticide application involves some really fascinating topics such as the fluid kinetics of droplet production, the use of additives (formulation) to impart certain physical properties to the spray and the engineering aspects of spray outlets (nozzles) and pressure sources (pumps). Much of this lies outside the scope of a 'Study in Biology', and readers are referred to METCALF and FLINT (1962) for a useful summary.

Certainly formulation (the solvent, carrier, additives used, etc.) and the method of application (see Fig. 7–1) can have almost greater influence on the efficiency and selectivity of kill than the choice of active ingredient. This is because it is still a long way in biological terms from the emission of a stream of pesticide to achieving kill of the pest.

The first problem with biological implications is to get the right amount of chemical on to the target substrate, which is usually the foliage of plants. As will be discussed later (section 1.3.3), a spray cloud is never uniform, but has a wide spectrum of droplet sizes. The larger droplets (although paradoxically forming the bulk of the spray volume) are too few in number to effect adequate coverage. If the leaf is at all wet or shiny, these large drops (e.g. 300 μm in diameter) will frequently just bounce off again (spray reflection). This reflection (Plate 1), and the effect thereon of different leaf surfaces, can be simply demonstrated by passing leaf discs laid on filter paper under a source of coloured drops (e.g. a burette containing a coloured solution and with the tap slightly open).

Small drops, although they provide potentially better coverage because of their enormous number, suffer from a rapid drop in momentum after they have left the spray nozzle. Even slight, almost imperceptible air movement will divert them away from the target (spray drift); moreover, small droplets tend to 'stream around' any obstacle to air flow (e.g. a leaf) and tend to accumulate, if at all, at the edges of leaves or on leaves which are edge-on to the spray. Such leaves may collect insecticide deposits seven times greater than leaves facing the spray.

When a droplet hits a surface it flattens. The more the surface tension properties of the spray and the characteristics of the surface resist the tendency of the drop to return to a sphere, the better will be the adhesion and 'spread' of the drop. Frequently 'spreaders' and 'wetters' are added to the spray to improve the 'flattening' of the drop, but excess additive of this kind will result in only a thin film of insecticide while most runs off on to the ground. The *lowest* amount of additive which gives 100 per cent wetting will provide the maximum insecticide deposit. Similarly, where oils are sprayed in water, the addition of emulsifiers may be needed to stop the emulsion 'breaking'. Once again the amount of additive is critical, because the aim is to deposit the oil on the leaf surface while the water carrier runs off. Excess emulsifier raises the proportion of oil in the run-off. The practical significance of some of these problems can be seen from a test where a spray of Bordeaux mixture (a fungicide) was applied to cocoa leaves to give a theoretical deposit of $25\ \mu g$ of copper per square millimetre of leaf. The maximum deposit achieved was $1.3\ \mu g\ mm^{-2}$ on the lower surface of the young leaves; on the waxy upper surface of mature leaves the deposit fell to $0.7\ \mu g\ mm^{-2}$.

Whatever deposit is finally achieved is immediately subject to erosion with time. Water evaporating from the leaf surface carries dissolved volatiles away, falling rain will wash deposit off the leaf and dust carried by the wind will abrade the insecticide layer. Leaf growth may create tensions in a brittle deposit so that it cracks and sloughs off; the deposit is also vulnerable to chemical degradation, both phytochemical oxidation and enzymatic within the plant or by the leaf surface microflora.

Even if an adequate toxic residue remains by the time the insect makes contact, such residue still has to accumulate in sufficient quantity at the internal site of action to exert its lethal effect on the insect.

The cuticle (external skeleton) of insects is not surprisingly an effective insulator preventing the movement of compounds between the external and internal environments. Insects, as terrestrial creatures, have to protect themselves against water loss, and so the cuticle is a potent barrier to the movement of water. This barrier is chiefly in the outer layers and particularly in the thin surface layer of wax. Insects allowed to move on an abrasive powder (e.g. carborundum) quickly abrade this wax and die of desiccation. The remainder of the cuticle is largely hydrophilic and resists the penetration of oils and waxes. In this way the cuticle can resist the penetration of compounds which are solely soluble in either water or oil. An important property of an insecticide is that it should have a good oil : water partition coefficient, and therefore be able to move in both phases and in this way through the cuticle. This property can often be established by a suitable choice of solvent.

The site of action of an insecticide is a specific tissue or even enzyme site in the insect and is a small fraction of the internal tissue surrounded by a mass of physiological and biochemical barriers or competitors for the toxic molecule. The rate of arrival of the toxin at this site will be the net result of penetration, toxication, detoxication, inert storage and excretion; all these will determine whether the amount reaching the site of action attains the threshold lethal dose. The main problems to be overcome within insects appear to be slow diffusion through the insect skin underlying the cuticle and through the other tissues, the lack of penetration of the sheathing of nerves (often the site of action) with a non-cellular sheath (the neural lamella) which is a barrier to ionized materials, the storage of toxin in fatty tissues and the breakdown of the toxin by insect enzymes.

1.3   The problems of pesticides

The wide-scale use of synthetic broad-spectrum pesticides, particularly certain organochlorines, resulted in certain obvious damaging side-effects. This caused CARSON (1962), an American journalist, to publicize the doubtful ethics of introducing large quantities of non-specific biocides into the environment. Her famous (or notorious, depending on personal opinion) book *The Silent Spring* grossly overstated the facts in such a way that politicians could not help but take notice of the resultant public outcry and pressure. Thus legislation and research funds were provided which the scientific community, already well aware of the problems, had not been able to secure with their more accurate and conservative memoranda through official channels. As *The Silent Spring* is such a landmark in the history of pest control, it is perhaps appropriate to discuss the problems of pesticides under Rachel Carson's own chapter headings. Organochlorines have all become tarred with the same brush, though it must be emphasized that they have varying properties and include some excellent chemicals (e.g. BHC) even from the environmental point of view.

1.3.1   Elixirs of death

Here Rachel Carson pointed out, as will already be evident from the description above of the modern synthetic insecticides, that the pesticides in regular use had very non-specific toxic mechanisms which rendered them poisonous to humans as well as to many other animals including the insects.

When such basically poisonous chemicals have to be handled, as in preparing and spraying the pesticide, the dangers to the farmer or grower do not need stressing. However, man accepts industrial risks in many of his occupations and there are established safety procedures for the handling of pesticides, just as for any other industrial activity. Many workers ignore the safety procedures and it is sometimes hard to blame them in this respect. For example, the farmer in the tropics would perhaps die of asphyxiation or heatstroke if he indeed donned 'fully protective clothing' while spraying a large area of crop.

Other fatalities result from ignorance or negligence—such as storing surplus insecticide in empty beer bottles. One has to remember that other toxic chemicals (aspirin, disinfectant, etc.) are regularly stored in the home and have been involved in similar fatalities.

Most concern must attach, not to such acute toxicity problems, but to any danger that lies in the chronic regular intake of small quantities of pesticide as residues on our food. Great alarm followed the discovery in the 1940s (based on post mortem analyses of accident victims and executed prisoners), that hardly a person could be found in the developed countries who did not carry measurable quantities of DDT in his body fat nor a restaurant meal clean of pesticide. This alarm was heightened by the realization that this DDT appeared to be retained in the body and that man showed a linear increase of body fat DDT year by year. However, as the years passed it became apparent that man attained a plateau level (usually around 10 ppm DDT in body fat) and that the amount did not then rise further. The question then remains 'Do such concentrations do us any harm?' It seems fair to conclude that at present there is little evidence either way. Carrying such compounds in our bodies is very unlikely to do us any good; of that we can probably be sure. Many workers have sought to show that insecticides marginally increase our risk to various terminal conditions such as heart disease and cancer; however, one must weigh such risks against what man has gained from being able to protect his crops at a time without effective alternatives.

1.3.2   And no birds sing

One of the most striking side-effects of organochlorine pesticides has been the death of many birds. This resulted partly from the widespread use of the compounds as seed dressings for grain which was eaten by birds, and partly because the pesticide applied to crops leached through the soil and finally arrived in waterways where it was concentrated in particular water layers and through animal food chains till the higher predators were affected. Predatory land birds such as eagles, kestrels or hawks as well as water fowl, particularly grebes, all suffered from acute organochlorine poisoning. Perhaps if Rachel Carson had titled this chapter 'And no worms turn' it would not have had the same impact on a human race peculiarly devoted to its 'feathered friends', yet it is among lower animals that the broad spectrum toxicity of pesticides has its most devastating repercussions.

Some of the repercussions of the toxicity of pesticides to the smaller non-target organisms are seen in the so-called 'resurgence' problems. A broad spectrum chemical may be highly effective against a highly mobile pest, but may also destroy the other organisms, including beneficial ones, on the crop. When the pest returns it is able to multiply without restraint from natural enemies, so that a far worse pest problem is created than was present before the pesticide was applied. Many examples of such an effect, including the classic one of para-oxon creating the 'most enormous cabbage aphid outbreak . . . ever . . . seen in England' are cited by RIPPER (1956). Often such spectacular resurgences are very short-term and easily reversible, but there are other examples which were slower, less spectacular and therefore much more permanent (long-term).

The use of pesticides may also lead to the appearance of new pests. The use of DDT against codling moth and of tar oils against overwintering eggs on apples promoted fruit tree red spider mite (*Panonychus ulmi*) from insignificance to a major pest. DDT killed its predators as well as stimulating its fecundity; tar oil killed the eggs of predators and competitors while red spider eggs selectively survived due to their breathing pores reaching above the oil film.

Two further points need making on the side-effects of pesticides on non-target organisms. Firstly, bees are important pollinators of many crops and are at considerable risk from insecticides. Although nearly all farmers and all growers avoid spraying when their own crops are in flower, bees may suffer from insecticide drift which can carry insecticide on to fodder crops and wild plants in the headlands and hedgerows.

Secondly, and included here for want of a more relevant heading, there is the effect of pesticides on the crops they are intended to protect. Sometimes the plant-poisoning effect (phytotoxicity) is spectacular as with the sudden leaf drop of sulphur-shy currant bushes when a sulphur containing pesticide or fungicide is used. Such dramatic phytotoxicity is usually spotted by the chemical manufacturer during the pesticide's development and the chemical will then be diverted to the herbicide section! Other more hidden expressions of phytotoxicity may only be discovered by chance research after the chemical has been in use for many years. Most pesticides cause some poisoning of the plant which is shown in a reduction in yield, sometimes as high as 10 or 15 per cent. Other examples are impaired crop flavour (taint), reduction of sweetness in fruits, reduction in fruit set and a tendency to accentuate the biennial cropping problem in apples. Phytotoxicity may also frequently result from formulation. For example, addition of high wetter concentrations will sometimes cause local damage where the 'run-off' drop has clung to a fruit and gradually concentrated by evaporation.

In conclusion, it must be pointed out that many of the various long-term problems of effects on non-target organisms (particularly the larger 'wild life') have now been reversed since the highly persistent organochlorines have mostly been replaced by the much shorter lived organophosphate pesticides.

1.3.3   Nature fights back

Probably the most serious problem of pesticides is that they can lose their effectiveness due to the 'appearance' of tolerant strains of the pest. The main reason is that pesticides create selection pressure on pest populations, which invariably have a genetic pool of widely differing susceptibility to the poison. Certain individuals have less permeable cuticles, faster storage of toxin in fat or are better equipped with enzyme systems for metabolizing the toxin. Such 'resistant' individuals survive the pesticide to breed the next generation (Fig. 1–2). There is also some circumstantial evidence that tolerance in pests may sometimes be due to a mutagenic effect of the pesticide extending the gene pool from which the selection takes place.

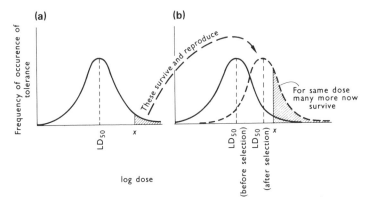

Fig. 1–2   Unnatural selection and the development of resistance. **(a)** Distribution of tolerances before selection. **(b)** Distribution of tolerances after selection compared with distribution before selection. *x*, the dose exerting selection pressure, kills fewer organisms in the population of progeny than in the population of parents. (From HASSALL, K. A., 1966, *Scient. Hort.,* **18**, 103–15; courtesy of author and the Horticultural Education Association.)

Two questions perhaps arise in the reader's mind: 'Do not natural enemies similarly become resistant?' and 'Why cannot we apply a dose of pesticide which leaves no survivors?'. As far as the first question is concerned, only very recently has an example of pesticide resistance in natural enemies been observed, in the case of a predatory orchard mite. A pesticide is likely to apply selection pressure to a large proportion of a pest population in any area and to be applied by a technique specifically designed to reach and make effective contact with the pest species. Natural enemies are behaviourally different and may often consume much untreated prey; they also tend to have fewer generations per year than the pest they attack. All this will tend to retard selection for

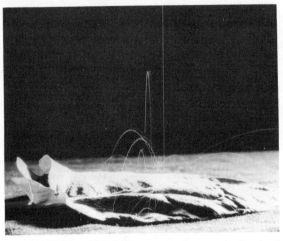

Plate 1 Photographic track of droplets impinging on a reflective leaf surface and 'bouncing' off instead of adhering. (Courtesy of DODD, G. D.)

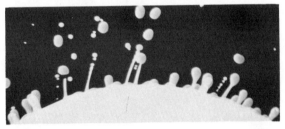

Plate 2 High speed photograph of spinning disc, showing the fragmentation of the liquid sheet into filaments, which rupture to produce main and satellite drops. (Courtesy of Plant Protection Ltd.)

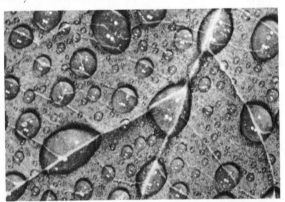

Plate 3 Close-up of part of a sprayed leaf to illustrate the distribution of different drop sizes in a spray. N.B. It is the smaller drops which provide the bulk of the spray coverage. (Courtesy of Plant Protection Ltd.)

pesticide tolerance. The orchard predatory mite and its prey (also a mite) overlap considerably spatially and behaviourally.

The answer to the second question is not that there are any individuals totally immune to the pesticide, but merely immune to the dosage that can safely be applied. There is often a relatively small margin of selectivity of action of the pesticide on insect and plant. Even where the margin is quite large, it cannot be utilized directly because of the way in which spray droplets are formed. As yet it is not possible to produce drops in sprays by a means other than what amounts to disintegrating the edge of a sheet of liquid. High speed photography (Plate 2) shows that when this happens, a filament supporting each main drop shatters to form several small (satellite) drops. The spray cloud therefore consists of very many small drops and fewer larger ones. The large drops contain most of the insecticide but account for only a small area of the crop surface covered by the spray (Plate 3). Too much insecticide in these drops, and the crop will be damaged. Because it is the small drops which provide the coverage, these must contain the effective pesticide dose. This can only be increased to a limit set by the phytotoxic properties of the large drops.

The 'tolerant pest strain' problem is so serious because there is a real danger that the appearance of such strains will outstrip the production of effective pesticides. As the problem is the inevitable consequence of pesticide use, it is not surprising that over half the world's pests show tolerance to at least one major group of insecticides. The production of new chemicals has never been a rapid process; a company may do well to market a new insecticide once every five years. Added to this, economic pressures are forcing several pesticide development companies to abandon this field of activity. The resistance problem is not yet so acute that new insecticides can command an astronomic price in competition with the older chemicals whose patents have expired and which can be manufactured by companies which do not carry the heavy overheads of development. Meanwhile, the costs of testing and developing a new pesticide have soared to approaching the £6 million mark. A new chemical must be protected by patent early in development, and it is not unusual for six of the fifteen years of patent life to expire before the product is marketed (Fig. 1–3). The chances of retrieving the development costs with an adequate profit in the short residual patent life are daily becoming slimmer, and it is in no way surprising that manufacturers are 'opting out'. The day can be envisaged when the development of new compounds may virtually have ceased before it receives a new stimulus from acute pesticide-tolerance problems the world over. No one would suggest that companies should pare down their procedures for testing pesticide safety, but it does seem vital that attention should quickly be given to extending patent life.

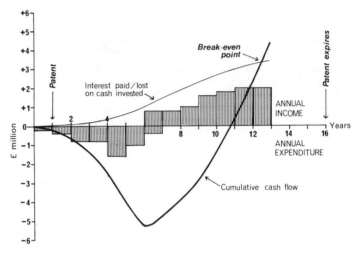

Fig. 1–3   Cash flow in the development of a hypothetical new insecticide, with a break-even position less than four years before expiry of the patent.

1.3.4   The other road

Rachel Carson denigrated the road via pesticides and suggested man should seek another road, particularly that to biological control of pests. The purpose of the succeeding chapters is to explore this other road, particularly to see how far a general alternative to chemical control emerges. Rachel Carson wondered what man's fate would be when he came to the end of the insecticide road, but for the actual event and its solution we must wait for the last chapter.

2 The Causes of Pest Outbreaks

2.1 The pest problem

Estimates of the pest problem on a world scale suggest that, without insect pests, world food production could be increased by about a third. As this represents the loss despite current control measures, it would clearly be catastrophic for mankind if control of insect pests were not attempted or should fail.

Obviously each insect individual has a fairly small food requirement. For example, a greenfly is unlikely to extract more than about 0.5 cm^3 of sap from a plant in its life-span; even a voracious caterpillar is likely only to consume 50 g (less than 2 oz) of the fresh weight of its host plant.

There must therefore be a great many insects to account for the damage they do, and insects can indeed be found in enormous numbers. There may often be 25 million per hectare of soil and 25 000 in flight over a hectare as compared with a human density over the dry land of the earth of about 0.14 per hectare. Obviously they are not all eroding man's food supply; however, numbers of just a single pest species per hectare of crop will often be comparable with such figures. One hectare of oats may harbour 22 million fly larvae and 222 million black bean aphids per hectare of sugar beet have been recorded. Both these infestation figures represent a rapid multiplication of the initial immigrants to annual crops, and indeed most insect species which cause pest problems have fantastic powers of increase. These powers, when expressed as statistics, would not seem out of place in a science fiction novel!

Such statistics are wildly unreal in ignoring, as they do, any restraints on population increase of mortality, food supply or space to live and in assuming optimum climatic conditions for the whole 12 months of a year! Yet they serve to illustrate the potential 'population explosion' rate that pest control must aim to suppress. Thus, the potential of cabbage aphid to have a new generation every two weeks and deposit 50 young is more dramatically (if nonsensically) expressed as the potential of one mother to produce in one year offspring weighing 250 million tonnes, encircling the equator nose-to-tail a million times. Equally startling is the notion that a pair of houseflies could cover the earth to a depth of 15 m with their offspring (200 million million individuals) in one year.

2.2   Insects outside the crop

If we look at uncultivated land, i.e. the roadside verges, hedgerows, commons and woodlands, it is striking that insect species do not seem to be found in very large numbers, the plants do not appear to suffer extensively from insect attack and numbers of insects do not fluctuate greatly from month to month or year to year.

It is therefore worth examining how the numbers of insects are regulated in nature. Clues can be obtained by studying populations in uncultivated land or by comparing populations in situations where the same insect is a pest and where it does not attain pest status. The regulation of animal numbers has been discussed by SOLOMON (1969) in this series; therefore, only a very brief discussion will be given here.

2.3   The regulation of insect numbers

The numbers of an insect species are clearly related to the ratio of births to deaths in a given time. Birth rate is influenced by many things, including weather, the food received by the insect and also nearly always by the degree of 'crowding' of the individuals (cf. territorial behaviour of birds). Death rate is influenced mainly by climate and natural enemies or disease; crowding may lead to emigration which, like death, leads to a reduction in the number of individuals in an area.

These influences can be classified under the two headings: 'Weather' and 'Competition' (Fig. 2–1): but more important for the present discussion is whether the process in which they are involved is 'regulatory' or 'non-regulatory'. Broadly speaking, a regulatory process must involve a positive *density-dependent* relationship between the

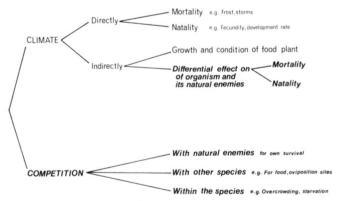

Fig. 2–1   Classification of environmental factors determining changes of abundance in the field. Density independent factors are shown in roman letters, density dependent factors in italics.

influence and the insect population, i.e. the restraint on the insect population rises *more than in proportion* to insect density as the latter increases.

This density relationship of restraints on populations can perhaps be explained most easily diagrammatically (Figs. 2–2, 2–3). It must also be pointed out that a relationship can be *inversely* density-dependent, i.e. the restraint on the population decreases as insect density rises (SOLOMON, 1969). For simplicity, inverse density-dependent relationships are omitted from this discussion.

Because insect populations tend to increase geometrically, a typical population increase can be represented as a straight line plot of the *logarithm* of density against time (Fig. 2–2a). If the restraint is very strong, the change in angle may be very great (Fig. 2–2c). These changes cannot be regarded as regulatory, for line (b) or (c), if extrapolated, will lead respectively to enormous populations and extinction.

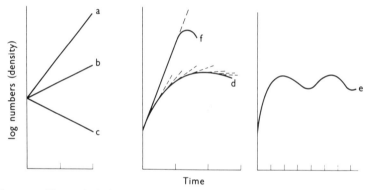

Fig. 2–2   Theoretical plots of log. insect numbers (density) against time under the influence of various factors (see text).

The effect of an influence with a density-dependent relationship to the population is quite different (e.g. many natural enemies). At increasing insect densities, the change in angle the influence produces will also increase (Fig 2–2d). Thus, the line will become a curve, and because the restraint decreases at lower insect densities, a 'regulated' oscillation of population density will result (Fig. 2–2e).

Some influences related to 'crowding' (e.g. starvation and emigration) become operative only at high densities, but then slow down the population increase very quickly (Fig. 2–2f).

Other 'crowding' phenomena exert their effects at much lower densities and represent subtle adaptations of insect populations for avoiding over-exploitation of their environment. Clearly natural selection will operate against the dangers to subsequent generations when individuals in an overpopulation situation have to face a 'scramble' type of competition for survival. Thus, competition for a convention ('conventional' competition) rather than for an absolute resource is a widespread phenomenon in animals, even among insects.

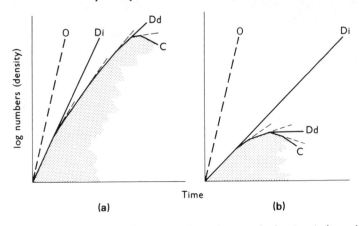

Fig. 2–3 The interplay of density dependent and density independent relationships to create high (a) or low (b) insect densities (see text).

Contacts between individuals may reduce fecundity, promote cannibalism and induce emigration or an arrest in reproduction.

We can represent two contrasting situations diagrammatically (Fig. 2–3):

(a) *Di*— Density-independent relationships favourable to the insect in relation to the optimum increase rate *(O)*;

   Dd—Density-dependent relationships produce a slow change in increase rate;

   C— Crowding influences appear late in time.

(b) *Di*— Density-independent relationships rather unfavourable to the insect;

   Dd—Density-dependent relationships produce a rapid change in increase rate;

   C— Crowding influences appear early in time.

As will be seen from the shaded areas of Fig. 2–3, situation (a) results in a population explosion; situation (b) results in population regulation.

2.4 Crops versus uncultivated land

It is tempting to relate situation (a) in Fig. 2–3 to crops (where insects are often found as pests) and (b) to uncultivated land (where insects rarely achieve pest status).

Although a huge over-simplification, the above relation is basically true and highlights sources of pest control other than toxic chemicals which are available to man and are discussed in the subsequent chapters.

Density-independent relationships: Under uncultivated conditions, the susceptibility to attack and quality of host plants is very variable—some plants are usually suitable for the pest at any one time, but at each time it is a different few (see Chapter 5). In the crop, the plants have been specially selected and managed so as to emerge, grow and mature together. Provided the pest arrives at the right time, it will find a very large proportion of the plants at a high level of suitability. Fertilizers, thinning, etc., are management practices which maintain a high plant quality suitable also for the rapid development of the infestation.

Density-dependent relationships: Natural enemies tend to be reduced within crops. There are four main reasons:

(i) Many pests have been introduced from abroad, and have been separated from their natural enemies which may not survive the new climate. The woolly aphid (*Eriosoma lanigerum*), for example, was imported into Britain from the U.S.A.; its effective wasp parasite (*Aphelinus mali*) is unable to survive the British winter.

(ii) The use of toxic chemicals depletes the natural enemy fauna of crops; they are often more susceptible to poisons than the pest and their re-invasion from other areas will only follow the re-appearance of the pest on the crop after a considerable interval.

(iii) The constant variety of prey and absence of fallow in uncultivated land make the latter a much more stable habitat for natural enemies than is the crop. A new crop must therefore be colonized by natural enemies from outside, and this may not occur till the prey is in short supply outside the crop or till pest numbers on the crop have built up to an attractive level.

(iv) Natural enemies may have requirements outside the crop for alternate prey or adult food (see Chapters 3 and 6).

Density-dependent relationships (crowding effects): In uncultivated land, the individual plants of any one kind are often few and scattered. It is hard for 'overcrowded' local populations to spread to other plants; emigration and starvation may therefore keep average densities in the area low. In the crop, infestation becomes spread much more evenly and average densities become high before emigration and other competition effects of crowding cause significant reductions.

2.5   An historical example

The pest population may therefore be compared to a motor car; slow progress can only be maintained uphill (in uncultivated land) by having the accelerator pushed to the floor-boards (phenomenal reproductive rate). The growing of crops bulldozes the hill away (plentiful food supply of high quality with reduced natural enemy populations) —without being able to help themselves, the insects zoom away at a terrifying speed! It is in fact noticeable that many pest insects occur 'naturally' at rather low population levels competing for specialized and ephemeral niches in uncultivated land; such species will have evolved particularly high reproductive rates as an adaptation to such an existence and to cope with the losses in transit of the highly mobile adults required for such life.

The ideas expressed in this chapter can be exemplified by the documented appearance of a new major pest in comparatively recent history.

In 1824 a British insect collector, Thomas Say, named an attractively striped beetle *Leptinotarsa decemlineata* as new to science. He had returned from an expedition to the Rocky Mountains, where he had found scattered individuals of the species on the eastern slopes, feeding on the weed buffalo-bur (in the potato family). When (30 years later), settlers brought potatoes as a crop to the region, the beetle spread eastwards as a causer of famine at a speed of 140 km a year, and soon reached Europe. Now this collector's rarity of Thomas Say, with the new name of Colorado beetle, is to be seen all over the world in illustrations on police station notice boards in the company of wanted criminals.

3   Biological Control

3.1   Introduction

The term 'biological control' is usually used for pest control by predators, parasites and pathogens. Many other controls are also 'biological', e.g. plant varieties, insect hormones, X-ray sterilization; however, the term 'biological control' has a long history of usage and would lose any meaning if taken too literally. This chapter deals with using predators and parasites for pest control (Plate 4); pathogens are discussed in the following chapter. HUFFAKER (1970) gives a comprehensive account of the subject. There is also a good introductory text by VAN DEN BOSCH and MESSENGER (1973).

The main advantages over chemical control claimed for biological control are:

(a) selectivity; pest problems are neither intensified nor new ones created;
(b) the beneficial organisms are already available—i.e. no manufacturing process is necessary;
(c) beneficial organisms can seek out and find the pest;
(d) beneficial organisms can increase in number and spread;
(e) the pest will be unable or slow to develop resistance to such control;
(f) the control is self-perpetuating.

Many of these claimed advantages can be challenged. For example, imported parasitic insects may occasionally carry a crop pathogen externally and thus bring a new problem into the crop. There have also been cases where an insect imported for biological control of an agricultural pest has subsequently attacked an insect introduced for the biological control of a weed.

Moreover, 'manufacturing' techniques are commonplace, not only with most pathogens, but also with predators and parasites when large numbers are required for release. Finally, although biological control can be self-perpetuating where it is non-exterminating, there is increased interest in using biological control in periodic applications rather like an insecticide, with pest and natural enemy dying out in consequence ('inundation'—3.2).

The main disadvantages of biological control are:

(a) control is slow;
(b) it is not exterminant, unless 'misused' (see above);
(c) it is often unpredictable;

(d) it is difficult and expensive to develop and apply;
(e) it requires expert supervision.

3.2   The techniques of biological control

(a) *Conservation*   Here the naturally occurring predators and parasites are encouraged by improving conditions for them, i.e. by cultivation procedures and by protecting them during pesticide applications. This approach is particularly suitable for the control of native pests in areas of mixed cultivation.

In California, it has proved possible to attract green lacewing (*Chrysopa* spp.) predators into crops *before* aphid populations are large by spraying the fields with a cheap protein hydrolysate mimicking the attraction of aphid honeydew to the predators.

(b) *Inoculation*   The natural enemy is liberated in relatively small numbers in the hope that it will establish itself. This approach is used particularly though not exclusively for the control of imported pests and a suitable natural enemy is sought in the country of origin of the pest species. The most suitable natural enemy may be relatively rare in the country of origin, but is the one which can find the pest at low population densities. The technique is most effective against sedentary pests of perennial crops or in 'ecological islands'.

A recent development in inoculation techniques is the even inoculation of both natural enemy and pest on new crops, particularly in glasshouses, to establish a biological control system of fast-breeding pests such as aphids and mites by the time the natural infestation arrives.

(c) *Inundation*   Large numbers of the natural enemy are reared in the laboratory and liberated on to the crop. Such releases are likely to lead to a fairly rapid control of the pest but will be followed by the disappearance of the natural enemies also. The technique is analogous to a pesticide application and is most widely used on annual crops.

3.3.   The history of biological control

As early as 1762, there is a record of the importation of predatory birds to Mauritius to control the red locust.

However, the startling success of biological control of *Icerya purchasi* (section 3.4) in 1889 is without doubt the real foundation of modern biological control. Biological control became the 'bandwaggon' of the first half of the 20th century, a bandwaggon which was only halted by the development of the cheap and effective synthetic insecticides. During this period, the cause of biological control suffered considerably at the hands of its own advocates, who had little knowledge of the causes of insect population change or of the pest characteristics or other conditions required for the success of biological control. Disillusionment about biological control prevailed widely from 1940 onwards. It is only recently that, largely due to the efforts of workers in Canada and California, biological control has been put on a sound ecological footing and has again become a respected philosophy of pest control.

The main argument in biological control today is whether multiple natural enemy importations are likely to prove more or less effective than a single introduction. To some extent the problem of competition between beneficial species is more likely to arise where such are already present before an importation than where an introduced pest becomes a problem in the virtual absence of natural controls; however, both ecological theory and practical experience suggest that multiple importations will rarely result in inferior control compared with the single release.

3.4   Some famous successes with biological control

(a) *Cottony cushion scale (Icerya purchasi)*   This famous project has already been briefly mentioned above. It is a story of human relationships just as much as of scientific achievement, and the account by DOUTT (1958) is well worth reading. In 1887 the new citrus industry was almost annihilated by the pest, and the federal entomologist C. V. Riley engaged Albert Koebele to investigate possible natural enemies of the pest in its country of origin, Australia. The first specimens of a ladybird (*Rodolia cardinalis*) (Plate 5) shipped from Australia to California arrived on the 30 November 1889. Within 15 months biological control of *Icerya* had been achieved at a cost of $1,500.

(b) *Coconut moth (Levuana iridescens)*   The biological control of this pest in Fiji in the late 1920s is notable in that it involved the parasite of a different genus of moth. The tachinid fly *Ptychomyia remota*, a parasite of *Artona catoxantha* in Malaya was established on *Levuana*, and 32 570 parasitized larvae were released. Within two years the coconut moth was under economic control over the whole island. This project was favoured in two ways. Firstly, the operation involved an island with a defined pest population; also, due to the mild climate, the pest generations overlapped continually and thus larvae were always available for oviposition by emerging flies.

(c) *Woolly aphid (Eriosoma lanigerum)*   This pest of apples entered Britain from the United States (hence its alternative name, American Blight) around 1785. Several attempts were made between 1924 and 1937 to establish the small parasitic wasp *Aphelinus mali* in Kent, but the introductions were not successful. In 1942 and 1943 it became more generally distributed and its control potential became apparent. Research showed that it could not survive the normally damp winter conditions in Britain. Biological control was maintained commercially thereafter by holding apple shoots with parasitized aphids over winter in

Plate 4 *(left)*   The parasitic wasp *Aphelinus* attacking the aphid *Aphis gossypii,* a pest of cucumbers. (Courtesy of Glasshouse Crops Research Institute.)
Plate 5 *(right)*   A famous success in biological control (see section 3.4). The vedalia beetle (*Rodolia cardinalis*) feeding on cottony cushion scale (*Icerya purchasi*). From DEBACH, 1964, p. 19 in *Biological Control of Insect Pests and Weeds,* Chapman and Hall, London.

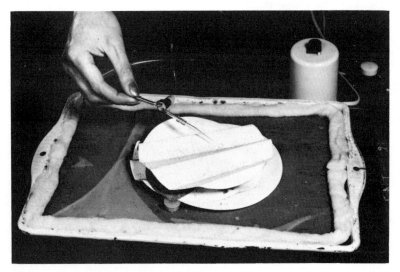

Plate 6   Rearing predators for biological control: dispensing *Phytoseiulus,* a mite predator of glasshouse red spider, from the culture into gelatin capsules for release in glasshouses. (Courtesy of Glasshouse Crops Research Institute.)

commercial cold stores. In spite of this cumbersome procedure, the method became commercially viable but was fairly soon abandoned after the advent of BHC, which gave efficient control of the pest.

(d) *Glasshouse whitefly (Trialeurodes vaporarium)*   In 1926, a Hertford-shire gardener noticed black whitefly scales on cucumber plants; these were identified at Cheshunt Experimental Station as parasitized by a small wasp, *Encarsia formosa*. The parasite produces several generations of parthenogenetic females in the summer, and breeds rapidly enough to exert control between 24–27°C. In 1929, the Empire Marketing Board financed the construction of a special glasshouse at Cheshunt to form an 'Encarsia factory', and in 1930 over a million parasitized scales on bundles of tomato leaves were supplied to nearly 60 growers. The method remained a commercially effective control till 1946 when the use of DDT against other glasshouse pests followed by the use of smoke generators in the early 1950s ended biological control by *Encarsia formosa* for nearly 20 years. However, the appearance in British glasshouses of strains of aphids resistant to insecticides has caused a new interest in *Encarsia* as part of a complete biological control system in glasshouse crops being developed at the Glasshouse Crops Research Institute.

(e) *Walnut aphid (Chromaphis juglandicola)*   In 1961, the parasitic wasp *Trioxys pallidus* was introduced to California from France. The introductions failed, although there was some establishment on the coastal plain of California outside the main walnut growing area in the Central Valley. There is a marked climatic contrast between the heat of the Great Central Valley and the mild coastal climate. A new ecotype of *Trioxys* was therefore introduced from the hot Central Plateau of Iran in 1968, and this release appears to have established successfully.

3.5 Failures with biological control

As indicated by the story of the walnut aphid above, a frequent cause of failure has been the different climate which the natural enemy experiences on importation. An almost identical story is the failure in California of the wasp *Aphytis lignanensis*, imported from South China in 1947 to control California red scale (*Aonidiella aurantii*). Although successful at the coast, the parasite failed to establish inland where the dry, hot summers and cold winters caused a high mortality.

Many natural enemies have a requirement for more than simply individuals of the pest species against which their use is contemplated:

(a) *Alternative hosts*   If the natural enemy is effective, the pest population may be reduced to a point where the natural enemy suffers from food shortage and dies or leaves the area. Such a case occurred in Puerto Rico, where a ladybird predator was imported against mulberry scale in 1938. The importation was initially highly successful, and the scale virtually disappeared. The ladybirds were last observed in

December 1943; in 1955 a heavy infestation by scale recurred. In some instances, therefore, so-called 'economically neutral' insects may be important in maintaining a biological control agent during periods of pest scarcity. Thus of 79 alternative hosts for Gypsy moth parasites listed in Germany, 45 occur in the ground cover rather than with the moth in the canopy of the forests.

(b) *Alternate hosts*   Sometimes a subsidiary host is an obligatory part of the life-history of a natural enemy. In Great Britain the diamond-back moth (*Plutella maculipennis*) is heavily parasitized by parasitic wasps, *Horogenes* spp. *Horogenes* emerges from the pest in the autumn, whereas the pest itself overwinters as a pupa and it is unsuitable for the overwintering generation of the parasite. It has been known since the 1930s that *Horogenes* must bridge the winter by a generation in some other caterpillar, but it was not till the 1950s that O. W. Richards located *Horogenes fenestralis* overwintering in a caterpillar (*Swammerdamia lutarea*) on hawthorn. *Swammerdamia* was till then just another 'economically-neutral' insect.

(c) *Flowers*   A glance along any British roadside in early summer will reveal very many insects feeding on the pollen and nectar of flowering plants, particularly hedge parsley and other Umbelliferae. These insects will mostly be females, feeding in this way to obtain protein for the yolk of their developing eggs. Many of the insects observed will be natural enemies of plant feeding insects and will include hover flies, small predatory flies and parasitic wasps, and dissection will reveal a high proportion of immature females.

Most agricultural crops are harvested before flowering or do not produce flowers early in the season; weeds within or outside the crop will therefore be the nearest source of flowers for newly emerged adult beneficial insects. At least one biological control project (WOLCOTT, 1942) has foundered because no flowers were available for the released parasite. Therefore the provision of flowers to aid biological control by native or imported beneficial insects is a frequently found recom-mendation.

3.6   The characteristics of effective natural enemies

Probably the major advance in biological control during the last decade has been a shift from empirical field assessment of the success of a 'hopeful' introduction to laboratory screening for effectiveness of potential introductions, particularly for the following ecological attributes:

(a) *Searching capacity*   If a pest is to be kept at low population density, the natural enemy must continue to search rather than emigrate from the area when its host becomes scarce.

(b) *Host specificity*   In general, host-specific natural enemies respond more precisely to changes in host density (section 2.3) than more polyphagous ones. Where, however, the pest population is periodically drastically reduced by other factors, such as harvest of the crop, a more general predator, which can maintain itself on other hosts at such times, may have the advantage.

(c) *Potential increase rate*   A short development time, large number of generations per year and high fecundity will be particularly useful attributes of a natural enemy to be used against a pest with similar properties, especially if the pest population fluctuates greatly under the influence of weather. Parthenogenesis (as found in *Encarsia*, the parasite of whitefly) gives the parasite a considerable numerical advantage, as only females of parasitic wasps directly perform biological control.

(d) *Climatic and niche adaptation*   The natural enemy should be able to survive in all the niches and throughout the climatic range occupied by the pest. The relationship of its development and voracity to temperature determines whether it can cause mortality sufficiently early in the pest annual cycle and whether it can avoid being 'outstripped' by the pest.

(e) *Ease of rearing*   For inundation and even inoculation procedures, it is useful if the natural enemy is easily cultured in the laboratory (Plate 6), perhaps on an easily cultured alternative prey or on an artificial food.

4 Newer Methods of Control with Biological Principles

4.1 Introduction

The problems of the continued widespread use of pesticides, and particularly the absorption of the words 'environmental pollution' into common vocabulary have caused scientists to look closely at any ideas for pest control which do not involve the widely used insect poisons. Biological control has been given a new searching look, but also novel ideas, crazy or otherwise at first glance, have been given serious consideration and even encouraged to develop in a way that would have been quite impossible twenty years ago.

4.2 Microbial pesticides

This is 'biological control' using insect diseases—a sort of 'germ warfare' against insects (BURGES and HUSSEY, 1971, is an excellent textbook on the subject). The most developed such 'pesticides' are virtually 'manufactured' and marketed as a spray or dispersible powder formulation. The main advantages claimed for microbial pesticides in contrast with toxic chemicals are that they:

(a) leave no toxic residues;
(b) have a high specifity for target organism;
(c) are compatible with toxic chemicals and can even sometimes be combined with them in a spray;
(d) have a low dosage requirement and are therefore cheap and versatile;
(e) lead to resistance in the target organism developing only slowly.

Set against these advantages are some very obvious disadvantages:

(a) careful timing of sprays is needed with those microbials which have an incubation period;
(b) specificity may be too great, so that certain life stages of the pest are virtually immune;
(c) each microbial has a pest population threshold below which the disease will not spread. Pest populations must be higher than this threshold if the microbial is to give effective control;
(d) the microbial may lose its virulence in the 'production' process;
(e) although not toxic to the consumer, the rotting cadavers of the diseased pests are an unacceptable additive to the produce;

(f) many microbials require damp climatic conditions for effective spread through the pest population;

(g) in contrast with natural enemies, microbials do not disperse on their own, but rely on the initial spray distribution and subsequently on the movements of their host for dissemination;

(h) the widespread distribution of large amounts of pathogen, when not too distantly related to diseases pathogenic to humans, is perhaps an unacceptable risk when examined against the high mutation rate of many of the organisms involved.

Disadvantage (c) has led to considerable emphasis of work in two areas where relatively high pest populations can be tolerated and where such populations frequently show natural disease infections. These are forest insects and cabbage caterpillars.

4.2.1 Contact microbials

Such infections penetrate the cuticle after deposition on the insect or contact by the insect with a treated surface. Penetration of surfaces is characteristic of fungi, and two genera of fungi have been particularly promising. These are species of *Beauveria* against caterpillars and *Entomophthora* against aphids and other small insects. Thus, for example, *Beauveria bassiana* has been tried against that very serious pest, the Colorado beetle. A problem with fungi is the need for a fairly high humidity before sporulation occurs on the corpses and the disease spreads further.

4.2.2 Ingested microbials

These are ingested by the pest and a drought-resistant stage of the microbial can be sprayed. In this way ingested microbials tend to be less humidity dependent than those which infect by contact.

4.2.2.1 VIRUS

The classic example of the use of virus to control a pest is the application of granulosis against the European pine sawfly in Canada. Granulosis viruses have been widely explored, but considerable success has also been achieved with polyhedrosis viruses (Plate 7). Viruses are applied in very dilute suspension, e.g. the contents of less than two diseased caterpillars per hectare in 300 l of water.

4.2.2.2 BACTERIA

Bacterial preparations are available as powders containing about 1000 resistant vegetative bodies per mg. These powders are wetted and sprayed, and the pest then may ingest vegetative bodies while feeding. Each vegetative body contains two structures, a spore and a protein crystal. When the vegetative body reaches the high pH of the gut, the protein crystal dissolves. This protein is the toxic principle which kills the pest; the spore is required for propagation of the disease.

High humidity is necessary for good control from bacterial disease but the preparations have great advantages in being harmless to honey-bees and being suitable for application close to harvest of edible foodstuffs. Thus there has been particular interest in using bacteria for the control of codling moth on apples. *Bacillus thuringiensis* is a very widely used bacterium.

A 'double microbial' has been tried in China to obtain improved codling moth control. Eelworms were infected with the bacterium and then sprayed. The infected nematodes eaten by the codling caterpillars pierce the gut and speed up the onset of bacterium multiplication in the dying grub.

4.2.2.3 NEMATODES

Although not strictly insect 'disease', nematodes have the character-istics of drought resistance and small size which render them amenable to storage and spray application in much the same way as diseases. Comparatively little use has been made of nematodes in this way, though they have been applied against Colorado beetle in Canada; they seem the only possible natural enemy against several important pests (i.e. mushroom fly, frit fly, thrips).

4.3 Repellants and attractants

4.3.1 Repellants

Chemical repellants have been investigated most fully with respect to mammals and birds. However, before effective insecticide alternatives became available, control of cabbage root fly (*Erioischia brassicae*) relied heavily on tarred discs placed around the bases of the plants to deter oviposition by the flies.

Repellant 'signals' of various kinds have been explored, particularly repellant sounds such as amplified bat calls to drive moths from orchards. It has also been found that the short wavelength light (sky) reflected from pieces of aluminium foil laid between the rows in a crop can greatly reduce the number of aphids landing on the plants by inducing descending individuals to fly upwards again. The material and labour costs of this method are high, but it has been used commercially in some high value crops (e.g. in the cut flower industry), where it has even given an effective control of aphid-borne virus disease.

4.3.2 Attractants

The use of attractants against insects has been developed much more than the use of repellants. Attractants can be combined with other control methods to introduce selectivity, i.e. the species to be controlled is selectively 'lured' to its doom! The high specificity of sex attractants has proved of particular value. Mating calls have been broadcast from traps but most promising is the development of attractive baits treated with insecticides and distributed within (or better still outside) the crop area. Very often only one sex (usually the male, which in many insects locates the female by scent) is affected, but this will reduce matings and can often give good control.

4.3.3   Antifeedants

The distinction between repelling an insect and inhibiting its feeding is that in the latter case the insects remain on the treated plants and starve to death rather than dispersing to seek food elsewhere. In contrast to many other control methods, therefore, some food for natural enemies remains for a time.

Antifeedants are found scattered in diverse groups of chemicals. Organotin compounds (e.g. triphenyltin) are perhaps particularly promising, and some triazenes and carbamates also show antifeedant properties. Present antifeedants appear to act on the taste receptors of the insect, and inhibit their perception of the stimuli to feed present in the host plant. As no systemic antifeedants have yet been discovered, sucking insects which pierce the treated surface are not affected, nor are new growth or patches of leaf missed by poor coverage protected.

4.4   Pheromones

It is becoming increasingly apparent that many behavioural activities of insects (e.g. movements, mating, aggregation) are under population control via chemical messengers produced by individuals in the population and liberated into the environment either as volatiles or in faeces, regurgitated food, etc. These chemicals have been given the general blanket term of 'pheromones' and they are available to man for manipulating insect behaviour either as extracts from insects or reproduced synthetically (the actual pheromone or a chemical 'mimic' thereof). Pheromones have particular advantages for pest control in that they are usually highly species specific, and are effective in very minute quantities. There are clearly occasions when the ability of man to make a pest react to a pheromone under his control has clear advantages. Thus sex attractants, as mentioned earlier, can be used to bring the pest to some other control measure (pesticide or chemosterilant); alternatively, it may become possible to use pheromones to aggregate a population in an area to be treated with pesticide.

However, the most promising and most general use for pheromones seems exactly in contrast to the obvious—i.e. not releasing but inhibiting certain pest reactions. Saturating the atmosphere with a pheromone (only minute quantities are needed) will do just this—the insects become habituated to the constant stimulus and the appropriate reactions (e.g. mating response to a sex pheromone) will be inhibited.

4.5   Genetic control

One of the advantages of genetic control is the '1 to many' principle—i.e. the 'treatment' of one individual which is then released alive may have a much greater influence on the population than killing it outright.

4.5.1   Radiation sterilization

The control of the screw-worm fly (*Callitroga* sp.) of cattle by radiation sterilization (KNIPLING, 1955) is one of the landmarks in the history of pest control. The release of males sterilized by a 5000 röntgen unit cobalt bomb on an island resulted in an eradication of the pest in eight weeks. Two things favoured success; firstly, the island had a finite fly population and, secondly, the insect is ideal in that the females only mate once. Therefore, a sterile mating is final. KNIPLING (1955) listed the theoretical requirements for best use of the technique:

(a) a method for the mass rearing of males;
(b) the released males must disperse rapidly throughout the native population;
(c) sterilization must not affect sexual competitiveness;
(d) preferably the females only mate once (e.g. *Callitroga*).
(e) the sterile males should be released when the native population is naturally low or has been reduced by insecticides.

Table 4–1 shows the rapid reduction of pest numbers that can occur if the population can be swamped by sterile males—this is of course the reason for applying the technique at a low pest population level.

However, the success story of *Callitroga* does not end there. The fly has now been virtually eliminated from the United States, following the erection of a factory on assembly line principles on a disused airfield. More than 50 tonnes of blood and meat go into the factory each week, to emerge as 100 million sterilized pupae which are air-lifted into the countryside in parachuted cardboard containers. The main difficulty has been to establish an 'island' situation in the face of continual invasion by wild flies across the Mexican border. Here sterile males are continually liberated in a 100-mile wide barrier zone.

4.5.2   Chemosterilants

In many ways, sterilization can be achieved more simply with chemicals than by exploding cobalt bombs, though most chemicals with insect sterilant properties have proved rather dangerous for man to handle. Most of the effective chemosterilants have proved to be either anti-metabolites (e.g. fluoracil) competing in enzyme systems, or alkylating agents (e.g. apholate, tepa) which affect alcohol radicals particularly in nucleic acid synthesis. The alkylating agents have been particularly successful in laboratory experiments with a wide range of pests. Although most chemosterilants seem particularly effective in male insects, there are also examples (e.g. amethopterine against houseflies or the fruit fly *Drosophila*) where the females appear the more susceptible sex.

Table 4–1 Theoretical population decline in each subsequent generation when a constant number of sterile males are released among a natural population of 1 million females and 1 million males. (From KNIPLING, E. F., 1955, *J. econ Ent.* **18**, 459.)

Generation	Number of virgin females in the area	Number of sterile males released each generation	Ratio of sterile to fertile males competing for each virgin female	Percentage of females mated to sterile males	Theoretical population of fertile females each subsequent generation
F₁	1 000 000	2 000 000	2 : 1	66.7	333 333
F₂	333 333	2 000 000	6 : 1	85.7	47 619
F₃	47 619	2 000 000	42 : 1	97.7	1 107
F₄	1 107	2 000 000	1807 : 1	99.95	Less than 1

Chemosterilants can, of course, be applied directly to crops or other substrates in the same way as pesticides. Just as with pesticides, however, a more restricted employment of the chemical seems desirable, such as a trap baited with the appropriate pest attractant (often a sex pheromone).

4.5.3. Hybrid sterility

Geographical races of the same insect may not always be genetically compatible. The possibility has been suggested of importing and releasing large numbers of males of an outside strain into a region once laboratory studies have indicated that such release would result in a large proportion of sterile matings. Analogous is the idea of genetically destroying the diapause adaptation to cold conditions of temperate pest populations. For example, the field cricket (*Teleogryllus commodus*) of Australia produces diapausing eggs in the colder south, but not in the north. Southern females produce only non-diapausing eggs when mated with northern males. Such matings following the release of northern males into the south should result in many eggs laid by the southern females failing to survive the winter.

4.6   Hormones (see section 4.4 for pheromones)

The idea that the internal chemical messengers of insects (particularly those which control moulting and metamorphosis) might find a use in pest control was proposed as long ago as 1958, when it was suggested that the spraying of moulting hormone should cause lethal premature moulting of pests, and also should be specific. In fact, internal hormones of insects are far from specific and interest in insect hormones for pest control has only recently been revived with the discovery that some plants appear to produce substances with insect hormone activity. Whether insects ever contact these substances is not certain, but the discovery has certainly stimulated interest in examining synthetic insect hormones and chemical 'hormone mimics' for their pest control potential. Interferences with development (both larval and embryonic) have been observed (Plate 8), but it seems likely that most promise lies in the approach where other chemicals or plant treatments are used to create a hormone imbalance in the insect indirectly through the host plant.

4.7   Competitive displacement

This is the concept of introducing a highly successful competitive species to displace and replace an existing species in an area. A plant pest would therefore usually need to be replaced by another plant feeder, and at first sight nothing would appear to have been gained. However, there are quite a number of advantageous possibilities which include:

(a) the newcomer need not be a potential pest if it outcompetes the resident species in some arena other than competition for food from the crop (e.g. in uncultivated land in the fallow season, for pupation sites);

(b) the newcomer, although a potential pest, is at least not a disease vector as the resident is, and can thus be more satisfactorily controlled;

(c) the newcomer may be severely affected by adverse climate (e.g. it is non-diapausing) in temperate conditions. The release of the newcomer, if it can displace the resident in one season, would then end in the disappearance of both species.

4.8 'Physicals'

The exploitation of physical properties of substances to combat pests is included here in the absence of any more appropriate section. Although they are 'newer methods', any 'biological principle' is rather tenuous.

Unsuccessful attempts have been made to desiccate fruit tree mites by spraying silica gel which absorbs the water-retaining fatty covering of the cuticle.

Sticky polybutene sprays to trap mites emerging from the eggs were actually marketed in the early '6os, particularly for use on cucumbers. Selling-points were that the mites would be unlikely to show resistance to this control measure and that larger predators would not be affected, but in practice the rather damaging effect to the plant of polybutene coated leaves caused the compounds to lose popularity.

4.9   Perspective

The proliferation of 'alternatives' to toxic chemicals reflects a 'liberation' of biologists in the field of pest control brought about by concern of governments and public about the side-effects of pesticides. A great number of useful techniques have emerged together with the danger that proponents of any one technique sometimes seek a generality for their interests comparable with pesticides. It would seem that, just as with classical biological control, an important practical step in the development of these methods is the identification of appropriate target pests.

Microbial pesticides, for example, have already played a large part in forestry, where pesticide sprays are costly in relation to profit and coverage is often difficult. Coupled with this, a fairly high pest population can be tolerated, humidity under the canopy remains high and favourable to the spread of diseases, also cadavers on the crop do not affect the sale of the product.

X-ray-sterilization and pheromone techniques have some clear specialized uses against native pests damaging crops at rather low densities and with appropriate behavioural characteristics.

Perhaps the most general 'newer' technique is the potential restriction of pesticide distribution afforded by the use of pest attractants. Chemosterilants perhaps also have a wider application in pest control, partly because the search for appropriate chemicals is not only of interest to the agricultural chemical industry but also because recognition of sterilization activity is not too costly an addition to the screening of many thousand new chemicals a year for traditional pesticide, fungicide or herbicide potential.

Plate 7 (*left*) Larvae of the European pine sawfly (*Neodiprion sertifer*) dying from infection with a nuclear polyhedrosis virus. (Courtesy of RIVERS, C. F.)
Plate 8 (*right*) Pupal deformity following application to large cabbage white butterfly (*Pieris brassicae*) caterpillars of a chemical with structural similarities to insect moulting hormone. (Courtesy of Plant Protection Ltd.)

Plate 9 Junction of experimental plots testing varietal resistance of lettuces to lettuce root aphid (*Pemphigus bursarius*). The impact of variety as a pest control measure is seen clearly in the contrast between the susceptible variety 'Mildura' in the foreground and the adjacent resistant variety 'Avoncrisp'. (Courtesy of DUNN, J. A.)

5   Varietal Control

5.1   Introduction

In many ways, the growing of crop varieties which are less attacked than others (Plate 9) or yield well in spite of attack is a very good pest control measure. Once such varieties are available, the control requires no extra labour and is therefore economical; moreover, the environment does not suffer from side-effects of the control measure. Until recently, 'resistant' plant varieties were sought energetically only in connection with problems where other means failed to control economically, and thus are widely available against plant diseases and plant parasitic eelworms. As the pressures for reducing pesticide applications currently increase, so varietal control of insect pests receives a new emphasis in pest control.

The subject of varietal control poses a problem of semantics. The words 'plant resistance' are often used as a synonym for varietal control; however, not all varieties which reduce losses from insect pests are 'resistant' in any normal sense. In this chapter, frequent use is made of the word resistance where this characteristic is implicit in the argument, otherwise the more general use of plant varieties in pest control will be termed varietal control or 'resistance' in quotes.

The plant breeder usually seeks genes for improved yield under exposure to pests in the wild parents of crop plants. At first sight it may seem surprising that man has not automatically selected for resistance in his crop plants by retaining the most productive types in the course of centuries. However, genetics is a science of this century only, and therefore most selection occurred before it was known that resistance may exist in an unproductive type and that crossing can often combine 'resistance' and productivity. Moreover, most early selection was for spot characters such as large fruit rather than productivity. Any resistance which was selected early on may have become nullified in the course of time by adapted races in the pest population or by recent importation of new pest species and races.

That varietal control involves the engineering of new or resurrected gene combinations by the plant breeder need not obscure the principle that the insects respond to phenomena more 'tangible' than the internal arrangement of plant nuclei. A variety is only effective through some chemical, physical or anatomical property of the gene combination. It is therefore possible to identify 'mechanisms' of varietal control and relate them to 'types' of control (i.e. the field expression of the mechanism on the pest population). This has been done in Fig. 5–1, where an attempt is included to arrange the mechanisms in order of stages in the pest infestation at which they appear most effective.

Many scientists have puzzled over the discrimination with which most plant-feeding insects select their hosts, and have argued that the apparent host selection is the result of what is really 'host avoidance'—i.e. plants have defences against most insects. SOUTHWOOD (1973) has concluded that plant feeding *per se* presents an insect with quite considerable difficulties and has also pointed out that plants will evolve to counteract damaging insect attack. Small airborne insects, which have little choice where they land, apparently reject and take-off again from plants we regard as their normal hosts with frequencies of anything between 50 and 95 per cent.

Several workers are currently querying whether the host restriction of many insects is not too strong to be explained in these terms alone and it is now being suggested that there exists a much more subtle relationship than 'conflict', involving avoidance by the insect of over-exploiting its host plant, the stabilization of insect populations and a maximization of genetic mixing of the insect's population throughout its range of distribution.

5.2   Classification of varietal control (Fig. 5–1)

The aim of varietal pest control is to reduce the losses in yield caused by pests. This is clearly achieved where no pest attack occurs, when the

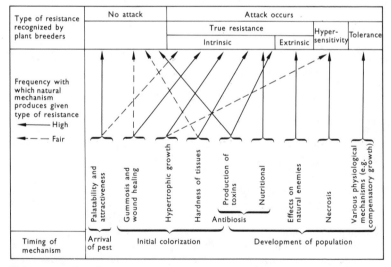

Fig. 5–1   A classification of plant resistance based on host plant influences on insect populations. (From VAN EMDEN, H. F., 1972, *SPAN,* **15,** 71–4; courtesy of Shell International Chemical Co. Ltd.)

variety succeeds in *escaping* the attack. More usually, a variety is attacked but may suffer less attack than the susceptible variety, because it is in some way *resistant*. This may be because of some *intrinsic* property of the plant, interacting directly with the pest species, or sometimes because of a more indirect *extrinsic* resistance operating through some other restraint on the pest (e.g. severe climate or natural enemies).

In contrast with such types of plants with apparent 'defences' against pest attack, improved yield may sometimes be gained from varieties showing a high degree of susceptibility. Some (*tolerant*) varieties seem able to yield well in spite of pest infestation and yet others are so susceptible as to be '*hypersensitive*' and collapse locally or entirely under attack; this collapse prevents the pest multiplying and spreading through the crop.

5.3 Mechanisms of 'resistance'

The term 'mechanism' is rather optimistic for our knowledge of the underlying causes of many instances of varietal control. Many examples, however, can be grouped under rather generalized headings (Fig. 5–1).

5.3.1 *Palatability and attractiveness*

However much opinions are divided on the evolutionary pressures which cause insects to show extreme discrimination, the actual discrimination plays into the hands of the applied entomologist who is trying to make varietal control effective. There is also general agreement on the actual plant characteristics for which the insects discriminate. Very often insects (e.g. the cabbage caterpillars) select plants which are botanically related and discriminate for characteristic secondary compounds (mustard oils in the above example). It can be shown very simply that cabbage caterpillars will not feed on leaves such as bean or lettuce, but will do so if the leaf surface is painted with mustard oils (rather dilute table mustard solution will do for this purpose). There is no evidence that these secondary substances play any role in the nutrition of an insect; many times it has been possible to 'wean' successive generations of a laboratory insect culture away from such substances.

It can also be shown that the same substances deter feeding by other insects, presumably the vast majority for any particular host plant.

However, insects do not select plants on these secondary substances alone, and many, particularly the more catholic feeders, also respond positively to adequate concentrations of particular nutrients, especially sugars. Insects may have further requirements for visual characteristics, particularly in terms of the wavelengths reflected from plant surfaces. Many Lepidoptera and Diptera will only oviposit where additionally leaf texture or the size of crevices (i.e. leaf axils) are suitable.

Red cabbage varieties are avoided by the cabbage aphid (*Brevicoryne brassicae*), although their nutrient content often seems more favourable for the aphids' multiplication than that of the green varieties which regularly suffer attack. The light green colour of 'Spanish White' onions seems to deter thrips from colonizing the plants, and apple sawflies seem to be attracted by a high ultra-violet reflection from apple flowers and therefore oviposit more on white-flowered varieties.

5.3.2   Gummosis and wound healing

Many plants protect themselves against wounding by rapid healing or by exuding gums, latexes and resins. In conifers, differences in resin flow have been implicated in the resistance of some pine species to attack by pine shoot moth (*Evetria buoliana*).

5.3.3   Plant growth form

Occasionally, features of external morphology are important in varietal control. Thus varieties of some solanaceous crops (tomatoes and potatoes) and beans possess large numbers of glandular hairs on the epidermis which make the plants resistant to aphids. These small pests break the hairs in walking over the plant surface; the broken hair exudes a fluid which adheres to the insects' legs and either fixes them to the plant or clogs the gripping structures so that the pest falls off the plant.

However, the very widespread involvement of plant growth form in varietal control is usually less obvious and is probably the least well exploited source of such control.

Rapid cutinization of epidermal cells and rapid cork formation in seedlings often protect fast maturing varieties from the many pests which concentrate their attack on young tissues. Thus differences in hardness of plant parts between varieties has been found important in varietal control of cabbage root fly (*Erioischia brassicae*) and wheat stem sawfly (*Cephus cinctus*). Many other insects attack the reproductive organs of plants and oviposit at flowering (e.g. pea moth)—varietal control involves selection for early flowering and a short flowering period.

5.3.4   Production of toxins

The production of what may loosely be referred to as 'toxins' by some plants has been closely examined in the development of varietal control of plant pathogenic fungi; phenols seem particularly widespread and useful fungistatic compounds in plants. They have also been implicated in the resistance of some apple varieties to root lesion nematode (*Pratylenchus penetrans*), and are probably produced as a defence in many plants in response to wounding (insect attack).

Little use has been made of toxins in the varietal control of insects, presumably because these pests are mobile and have evolved to recognize and avoid such unfavourable plants. It is also possible that 'toxins' are involved in some examples of escape or resistance at present ascribed to low palatability and attractiveness.

There is some evidence that certain pigments in cotton plants (gossypol and quercetin) may be toxic to the bollworm (*Heliothis zea*) and that varieties bred with an increased concentration of these pigments might prove resistant to the pest.

5.3.5   Nutrition

In spite of a great deal of knowledge on insect nutrition and how this can be varied through the host plant, little is known of its importance in influencing the resistance of crop varieties to pests. Some information is, however, available on aphids, which mainly feed in the phloem. Phloem contents can be altered without necessarily affecting the value of the plant for man and his domestic animals. Extensive work in Canada has linked the susceptibility of pea varieties to the pea aphid (*Acyrthosiphon pisum*) with the soluble nitrogen content of the plant, particularly the amino acids. 'Perfection' gave much denser amino-acid chromatograms than 'Champion' and on another resistant variety ('Onward') the aphids were comparable with aphids fed on a susceptible variety but starved for 10 hours daily. Recent work at Reading has linked resistance of brassicas versus two aphids, the cabbage aphid (*Brevicoryne brassicae*) and the peach potato aphid (*Myzus persicae*), to changes in the amino-acid spectrum. Different amino acids affected the two species and, whilst some amino acids were favourable to aphid increase, others were unfavourable.

5.3.6   Effects on natural enemies

Such effects result in extrinsic resistance, but few examples are known. It is probable that current techniques of screening for plant resistance would not detect potentials for extrinsic resistance.

Probably the best known example is that of open-leaved crucifer varieties which make it so much easier for parasites to find their cabbage caterpillar hosts that a high rate of parasitization results where open-leaved varieties are grown.

5.3.7   Tolerance

Tolerant varieties offer an ideal pest control; neither the grower nor the pest is under pressure to 'react' with either pesticides or 'adaptation' respectively. However, varieties able to withstand normal pest attack without a diminution of growth or yield are rare indeed, and often the underlying mechanism is not understood. Such is the case in the classic example of tolerance, the tea varieties tolerant to shot-hole borer (*Xyleborus fornicatus*). The most common tolerance phenomenon is to pests attacking plant roots and here tolerance depends on the ability of a variety to produce new growth which compensates for root grazing by the pests but is not an undue drain on the nutrient supply to the aerial parts of the plant.

5.3.8 Necrosis

Necrosis is a mechanism of 'resistance' involving hypersensitivity—an apparent contradiction in terms. The plant is so sensitive to attack that immediate death of the affected tissues or of the entire plant ensues. At first sight this appears disadvantageous to the grower in terms of yield, but of course this need not be so. The development of the attack, the multiplication of the organism and its spread through the stand may well be inhibited or limited. Necrosis is also a major source of resistance to invasion by fungal parasites and nematodes. One of the few examples of plant necrosis in 'resistance' to insects concerns the woolly aphid (*Eriosoma lanigerum*). Here the insect requires the apple plant to react to attack by producing new undifferentiated callus parenchyma on which the pest then feeds, but some apple varieties react instead with a rapid protective necrosis and the pest is denied a suitable substrate for its multiplication.

5.4   The use of resistant varieties

Readers are recommended to consult PAINTER'S (1951) treatise on 'Insect Resistance in Crop Plants', but some of the problems in utilizing such an apparently 'ideal' form of pest control are briefly discussed below.

5.4.1   The difficulties of utilizing pest resistant plant varieties

(a) *Variability between pests*   It commonly happens that resistance to organism A is linked with susceptibility to organism B. Commonly resistance to pests (often related to a high carbohydrate/nitrogen ratio in the foliage) lowers the defences of the plant to attack by fungal pathogens which are favoured by such a ratio. Thus one student practical class in the Horticulture Department of Reading University ranks apple varieties in order of resistance to red spider mite and exactly the reciprocal order is obtained with another class examining the distribution of apple mildew disease! Obviously varieties must be tested in field trials out of doors exposed to all expected and potential harmful organisms in as many different areas as possible.

(b) *'Breakdown' of resistance*   Resistance of a plant variety is no more a permanent control of an insect pest than is an individual pesticide. Both control measures exert a selection pressure on the pest and the minority strain not affected by the control measure will become more common. The only difference is that this may happen more rapidly with a pesticide (see section 1.3.3). Plant resistance has not yet been used extensively against pests, so no clear cases of a 'breakdown' of resistance with time can be quoted, but the phenomenon is well known to plant pathologists (e.g. cereal rusts) and to nematologists.

An analogous problem is that races or 'biotypes' of pests vary in different regions. Pea varieties tested against pea aphid in Canada and Kansas showed a quite differing order of resistance in the two areas. Biotype exchange showed that the differences were not environmental, but related to the particular strains of the aphid occurring in the two areas.

(c) *Environmental factors* Plant 'resistance' is the result of an interaction of insect behaviour and physiology with definite plant characteristics. In as much as the resistance characteristics are environmentally variable, so climate or soil type of an area may affect them so as to render the plant susceptible.

Because of the environmental variability of plant resistance, tests should be carried out in as many seasons and also climatically or topographically different areas as possible.

5.4.2   *Maintaining the effectiveness of resistant plant varieties*

(a) Where seed is collected in the field it is obviously essential to prevent pollen from other varieties fertilizing the ovaries and thus destroying the genetic background for resistance.

(b) The development of isogenetic lines. Isogenetic lines involve a series of selections incorporating identical agricultural or horticultural characters of the crop, but differing in their resistance to strains of a pest. Different selections from these lines can be chosen for the pest races and other conditions current in a particular area at a given time.

(c) Strains of the pest tolerant to the plant's resistance mechanism are likely to be selected with the greatest rapidity where the variety at first shows immunity to the pest. Thus it is in many ways more profitable to use lower degrees of plant resistance either in combination with other elements of an integrated central programme (see Chapter 7) or with each other. Several 'slight' resistance mechanisms add up in effect geometrically rather than arithmetically and can be surprisingly efficient and flexible as well as 'checkmating' shifts in the races of pests colonizing the crop.

6 Cultural, Physical and Legislative Controls

6.1 Cultural control

Before the advent of the modern synthetic insecticides, man's chief weapon against insects was to try to disrupt their life cycles by periodically denying them their food plant and to achieve the maximum control that a manipulation of ordinary agricultural practices would allow. Such 'cultural control' measures could not match the control that the new insecticides achieved, and many fell into abeyance, only currently to be resurrected as part of multiple control programmes (Chapter 7).

6.1.1 Cultivation of soil

Many insects live or hibernate in the soil, selecting suitable temperature and humidity conditions. These conditions can be disturbed by ploughing, which creates temporary drought conditions in the upper soil layers and may even expose larvae and pupae to the full radiation of the sun. Many of these insects will be eaten by birds, and pigs have even been brought on to ploughed fields for the especial purpose of picking out and eating white-grubs (beetle larvae pests of cereals). Other pupae and eggs may be buried by ploughing to a depth from which they fail to reach the soil surface after emerging. Yet other individuals will be killed mechanically by rough contact with soil clumps, and root aphids (e.g. cereal root aphid) depressed by the break-up of the ant colonies which tend them.

6.1.2 Clean cultivation

Standard farm hygiene often has a pest control purpose. The destruction of crop residues removes residual pest populations (e.g. pea moth pupae in unharvested peas, stalk-boring grubs in cereal stubble), and eliminates plant debris on the soil surface in which many pests find shelter for hibernation. The elimination of weeds deprives the pest of what may be essential alternative food if the insect has a longer feeding period than the crop season. Another aspect of clean cultivation is 'rogueing'—the removal and destruction of infected material where there is a danger of spread to other parts of the crop.

6.1.3 Manuring

Rapid healthy plant growth generally compensates for some damage by pests; weak, deprived plants may easily be killed out by equivalent attack. For example, good root systems will clearly withstand root grazing by pests where weak root systems will not. Similarly, shot-hole borer (*Xyleborus fornicatus*) damage on tea in Ceylon was successfully reduced by fertilizing the bushes with nitrogen. The stimulation in growth enables the bush to form new tissue over the beetle gallery entrances so that abcission of the branches at the gallery no longer occurs.

However, just as fertilizer produces a more nutritious plant for man, so many insects also benefit. Aphids, leafhoppers, mites, thrips and leaf mining grubs have all been found to breed or develop more rapidly on plants given good nitrogen fertilization. With aphids, for example, it is however possible to manipulate manuring to general advantage. Aphids are sap and not leaf tissue feeders, and good potassium fertilization will reduce nitrogen available in the sap without impairing the value of the leaf protein. Here is a field ripe for useful and relatively simple experimentation—how far can we 'induce' plant resistance (Chapter 5) by physiological treatments to plants?

6.1.4   Irrigation

Many workers ascribe pest control effects to excessive irrigation or watering: pests may be washed off the plants, drowned or suffocated, soil insects may be killed by the pressure of colloidal particles in saturated soil, or there may be some physiological change in the plant inimical to the pest (e.g. aphids, which tend to do badly on well irrigated plants and benefit from periodic wilting of the plant). Flooding vineyards was found to be effective in the control of *Phylloxera* (an aphid-like insect) and it has been used at times in spite of the harm it does to the soil.

Where, however, irrigation is used so that a crop may be grown at all, as in California and large areas of the Middle West, pests may seek out the only lush vegetation (i.e. the crop) in an area. It is, for example, a well-known phenomenon in some arid areas, that pest incidence on cotton rises dramatically following irrigation. In California and Peru, however, irrigation has also enabled a pest-natural enemy complex to persist and reduce the importance of bollworms as pests.

6.1.5   Strip farming

In the days before intensive agriculture, strip farming was a normal practice and had two main beneficial effects in terms of pest avoidance. Firstly, an intervening strip of non-suitable food prevented movement of pests from one strip of a crop to another or from one suitable crop to a different one. Also, adjacent strips shared unspecialized natural enemies which could move when food (pests) built up on the neighbouring strips. The abandonment of strip farming in Peru some 30 years ago has been given as the reason for bollworm outbreak on cotton there, and certainly re-diversifying the cotton agro-ecosystem (Plate 11) has now greatly reduced the incidence of the pest. The choice of adjacent crops is, of course, more important than the simple decision to diversify. Juxtaposition of wheat and maize, for example, would actually intensify problems of shared pests such as chinch bugs and eelworms, whereas separating the crops by a strip planted to potatoes would reduce pest damage on the cereal crops. The management of strips of crops compared with large monocultures is a serious disadvantage to the method, and it is only likely to be introduced as a 'last resort' in intensively farmed areas. Moreover, some pests (e.g. some grasshoppers) lay eggs at the edges of crops and can become a serious problem when, as with strip farming, the edge forms a large proportion of the crop!

6.1.6   Crop rotation and isolation

Attempting to separate the pest from its host plant in time or space is one of the oldest and most widespread farm practices often directly determined by pest problems, and is still one of the most effective controls of some eelworm problems. Crop rotation normally reduces and delays attack rather than giving complete control because, although it has importance within a given field, it is a less effective restraint over an area as a whole. Most pests have strong migratory powers or, if not, can frequently survive rotation on wild host plants (see section 6.2.1). Moreover, crop rotation normally means that a particular crop is always grown somewhere in the area. Thus the common rotation within a field of grasses or cereals, legumes and root crops does not result in the absence of any of these crops on the farm as a whole. Rotation is most effective against soil pests (e.g. white grubs and wireworms) which take several years to reach maturity. Crop rotation relies on the fact that there are usually only a few general feeders among the pests found across the rotation. For example, of fifty serious insect pests of the maize, wheat and red clover rotation, only three are important pests of all three crops.

Attempts to avoid pests by isolating crops from regularly infested sites are frequently designed to prevent insect-borne diseases from reaching the crop. Because wild plants (see section 6.2.1) form reservoirs of both the insect vectors and the diseases they carry, the method has rarely proved successful on a regional scale.

6.1.7   Trap crops

If insect pests can be concentrated onto particular small areas of a field, they can then be destroyed with locally applied insecticide or by ploughing in or otherwise destroying the vegetation (e.g. feeding to livestock). Such concentration of pests may be induced by position (e.g. edge rows for swede-midge; height differential of trap plants and the crop; relation to windbreaks (see section 6.2.2); earlier sowing (e.g. corn ear worm); spraying with attractants or choosing a specially attractive plant as the trap crop (e.g. kale for certain bug pests of cabbages). A specially interesting example of a trap crop is the use in Canada some 20 years ago of a non-crop trap plant (brome grass) planted in a 15–20 m strip around wheat fields to control the stem-boring sawfly *Cephus cinctus*. The adults did not penetrate into the wheat crop but laid eggs on the brome grass in which many larvae developed per stem. It was not necessary or even advisable to destroy the grass, for the grubs cannibalized one another and even most of the eventual survivors failed to survive in the grass although their parasites were able to emerge. Thus the brome grass produced parasites but very few sawflies.

6.1.8 Sowing and harvest practices

Variations of sowing date can control pests, most of which show some seasonal frequency, either by the crop avoiding the egg-laying period of the pest or by allowing the plants to have aged to a resistant stage by the time the pest appears. Just as one example, Hessian fly (*Mayetiola destructor*), has a predictable flight peak of limited duration and a few days' delay in sowing wheat can make all the difference between a good and bad crop.

Another possibility is to increase the seed rate to compensate for the expected plant loss from pests such as flea beetles, seed fly or stem borers. Increasing seed rate may also reduce the infestation of many airborne pests (e.g. aphids and small flies) which seem to find dense plantings less attractive.

Early harvesting removes pests (especially cereal pests) from the field before they can emerge and perpetuate the population in the area. Even if the pests then emerge in store, they are either easily killed there or perish in the absence of a fresh plant substrate.

6.2 The importance of non-crop plants in crop pest problems

Enough has been said about strip farming and the problems that alternative plant hosts present for crop rotation to indicate that other plants, particularly 'wild' flora, in the agro-ecosystem are of considerable importance. This subject has been discussed in more detail by VAN EMDEN (1965) and LEWIS (1965a), and only a brief summary will be given here.

It is possible to distinguish biological and physical components in the relationship of pests to crops and non-crop plants. Biological components concern the actual plant species of non-crop plants; physical components, on the other hand, could persist if the plants were replaced by artificial structures (e.g. a wooden fence in lieu of a hedge).

6.2.1   Biological components (Fig. 6–1)

Non-crop plants as weeds or present in adjacent uncultivated land may have an important place in the life-history and biology of insects found on the adjacent crop (1). Most insect pests also feed on 'wild' plants, especially if these are botanically related (1a) to the crop; the pests may use such plants as food if the crop season is shorter than the insect feeding season (1c). Also, irregular events such as the use of weedkillers on roadside verges may 'drive' the insect onto the crop (1c). Non-crop plants can therefore form a reservoir of crop pests which will also maintain the species in an area if the crop is absent because of rotation. In addition to forming a reservoir of pests, non-crop plants form an important reservoir of crop diseases from which the new crop can be infected each year. Very often infected wild plants show no symptoms of disease, and its presence can only be shown by electron microscopy or the infection of a crop plant when the vectoring insect is transferred.

Wild plants often provide different nutrition from crop plants, and pests arriving on crops from outside vegetation may differ in insecticide resistance and fertility (usually both to man's disadvantage) from ones reared on the crop (1b).

Flowering weeds in or outside the crop are frequently the only source of flowers in the agro-ecosystem for the many insects which feed on pollen and nectar as adults before they can mature their eggs. Although several pest insects fall into this category (2) (e.g. cabbage root fly), most importance is probably attached to the beneficial impact of such flowers in maintaining biological control (5) (see also section 3.5).

For the same reasons that pests utilize plants outside the crop for food, phytophagous insects such plants support may be valuable or even essential for maintaining natural enemies in an area (3) (see section 2.4 for fuller discussion). Some general predators (e.g. some bugs) may even turn to plant feeding when animal prey is scarce (4). As for pest insects, so natural enemies reared outside the crop may differ in biological characteristics from those reared in the crop (3e). Little is known about such important matters, yet they are amenable to quite simple experimentation.

6.2.2   Physical components (Fig. 6–2)

The interference to air currents caused by an upstanding barrier (e.g. a hedge) causes turbulence on the lee side and small airborne insects (many of which are pests) are deposited on the crop by the down-currents so caused (1). There is also some deposition of insects close to the windward side of the hedge. LEWIS (1965b) demonstrated these effects brilliantly with lettuces and an artificial windbreak—the root aphid infested lettuces died and the dead plants stood out as brown

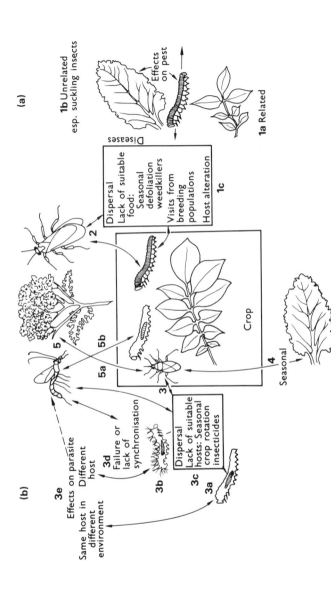

Fig. 6-1 Biological relationships between uncultivated land and both pest (**a**) and beneficial (**b**) insects. Injurious species feed on both crop and related (1a) or unrelated (1b) wild plants for a variety of reasons (1c); Pest insects (2), predators (5a) and parasites (5b) may also feed at flowers (5). (3) Beneficial insects may utilize alternative prey related (3a) or unrelated (3b) to the crop pests, for several reasons (3c, d, and such alternative food may effect the physiology of the parasite (3e). Sometimes, carnivorous insects feed phytophagously on uncultivated land (4). (From VAN EMDEN, H. F. 1965; courtesy of the Horticultural Education Association.)

bands showing the exact areas of deposition (Plate 10); the same effect can be seen in the deposition of snow around a windbreak. Such deposition of disease-carrying aphids accounts for the 'bands' of yellows virus infected beet and potatoes frequently noticed at the edges of the fields.

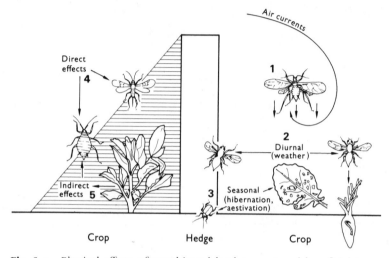

Fig. 6–2   Physical effects of uncultivated land on pest and beneficial insects. Such land may interfere with the transport of insects in air-currents (1) or provide diurnal (2) or seasonal (3) shelter. The microclimatic conditions adjacent to a hedge may influence the activity of insects (4) or indirectly affect them (5) via effects on the host plant.

Other edge-effects of noticeable pest damage occur because of movement (2) between the crop and the physical shelter provided by hedges either against the heat and low humidity of the day (e.g. carrot root fly) or against extreme heat (summer) or cold (winter), when insects seek the shelter of debris and fallen leaves in the hedge (3) for overwintering (e.g. flea beetles) or aestivation (e.g. blossom weevils).

A further physical impact hedges may have is through the shade and shelter they give over part of the crop area. This impact may be direct (4)—for example, most insect parasites fly and seek prey more actively in the sun than the shade, or indirect (5)—in that shaded plants usually provide poorer nutrition for pests than those in the open.

6.2.3   Conclusions on non-crop plants

Non-crop plants clearly have both debit and credit properties with respect to pest control. It is, however, essential to bear in mind that they have certain essential irreplaceable properties, particularly in maintaining a reservoir of general natural enemies. Such structural diversity of the agro-ecosystem allows natural enemies, based on non-pest species in the hedgerow, to forge trophic links with pests on the crop. There are probably many examples to be discovered similar to that of *Horogenes* parasitizing *Plutella* (see section 3.5), waiting for man to break the important trophic link. Hedges and similar refuges of floral diversity in agriculture may have great nuisance value in pest as well as in other terms, but it is important that we retain some 'weed' outside the crop until we are sure that valuable components can be replaced once removed or, if we are foolhardy, we abandon seeking biological components in pest control.

6.3   Physical controls

Such controls aim to reduce pest populations by using devices which affect them directly or alter their physical environment. They may be hard to distinguish from cultural controls and are frequently labour-intensive. For example, in the early days of pest control in under-developed countries, hand picking and foot-crushing of larger pests (e.g. caterpillars) was an economically viable and effective control. Grease bands around the trunks of apple trees to trap the ascending flightless females of winter moth were standard practice for many years before the advent of modern insecticides.

Most laborious practices have now proved too expensive and rather sophisticated machinery now represents 'physical controls', though as yet it has still to leave the research bench. Here we can cite as examples traps with fans to smash entering pests and the amplification of attractive or repellent sounds respectively to attract pests to some other doom or to drive them from the crops (section 4.3).

The only method in this category which has really stood the test of time is the hot water treatment of plant storage organs (e.g. roots, corms and bulbs) to kill concealed pests such as bulb flies and eelworms. Unfortunately, generality of the technique is limited by the thermal death point of many pests being quite close to that which already damages the plant organ.

6.4   Legislative controls

In many ways this is not a distinct control method but merely represents the legal enforcement of other control measures.

6.4.1   *Quarantine*

Most countries operate quarantine laws to allow inspection at the point of entry of all produce, etc., which might harbour foreign pests and to enforce strict isolation of any species imported for study (e.g. for biological control research).

6.4.2   Eradication

Particularly serious pests may be subject to a 'notification order', whereby any farmer who suspects the pest may have appeared on his crop must notify the appropriate authorities, who then undertake pest eradication.

6.4.3   Certification

Certain plants, seeds, tubers etc. subject to particular pests or diseases may not be sold unless free of the problems. For example, 'The Sale of Diseased Plants Order' (a series of laws passed in Britain between 1927 and 1952) prohibits the sale of plants with infestations of several pests, including glasshouse whitefly, but has probably been honoured more in the breach than in the observance.

6.4.4   Rotation orders

Rotation is among cultural practices which have been subject to legal enforcement in various countries at various times (e.g. sugar beet rotation to control beet eelworm in Britain).

7 Multiple Control Programmes

7.1 Introduction

It was mentioned in Chapter 1 that man has found situations where the insecticide 'road' has run out. Insecticides have failed because of tolerant pest strains, and this first happened in the mid and late 1950s on cotton in Peru, on alfalfa in California and chrysanthemums under glass in Britain.

Rachel CARSON (1962) had advocated that man must choose between chemical and biological control; he (man) was 'standing at a fork' of the ways. The first sentence of her final chapter ('The Other Road') in fact begins with the sentence 'We stand now where two roads diverge'. In the light of what was already happening at that time on alfalfa in California, and what has happened since elsewhere, the sentence stands out as perhaps the most interestingly misled ever written about pest control.

When insecticides failed, 'insecticide' man was not brought to realize that 'biological control' man was on the right road after all. Rachel Carson's 'fork' of the roads was really only mythical; in the event 'insecticide' man and 'biological control' man, we may guess perhaps to each other's surprise, came together where the two converging, not diverging, roads finally met—the development of 'integrated' control (STERN, SMITH, VAN DEN BOSCH and HAGEN, 1959).

7.2 Classic examples of integrated control

7.2.1 Cotton pests in the Cañete valley of Peru

The cotton crop in this valley was predisposed to ecological disaster once broad spectrum insecticides were applied, because it was an irrigated monoculture in an otherwise dry and biologically impoverished area. Thus an entire ecosystem and its entire faunal complement was being blanketed with pesticide. Rapid and devastating problems, including most of the side effects of 1.3 occurred as early as 1955. In 1956, legislation based on cultural practices (Plate 11), the re-population of the valley with beneficial insects and a return to older, more selective insecticides combined to reduce pest problems and so raise cotton yields dramatically in the late '50s.

7.2.2 Spotted alfalfa aphid in California

The spotted alfalfa aphid (*Therioaphis trifolii*) was first seen in California in 1954. In the late '50s the aphid developed resistance to organophosphate insecticides and crop losses became critical. The

Plate 10 Damage to lettuces by the lettuce root aphid (*Pemphigus bursarius*) illustrating the localized deposition of the immigrating aphids in the shelter of a 45 per cent permeable fence. During the immigration, the prevailing wind was approximately right to left across the picture. The smaller windward and larger leeward areas of shelter are shown by the bare areas where lettuces have been killed out by the aphids. (Courtesy of LEWIS, T.)

Plate 11 Mixed cultivation in the Cañete valley of Peru: part of a classic integrated control programme against cotton pests. (Courtesy of *The Furrow*, JOHN DEERE.)

courageous step was taken of applying an organophosphate (demeton) at low dose! Some aphids were killed; of course many survived, but so did many of the natural enemies which had not been effective controls on their own but were now able to control the surviving aphids. Within one year of applying the integrated control programme, the crisis was over.

7.2.3   Peach potato aphid (Myzus persicae) on chrysanthemums under glass in Britain

The advent of organophosphate resistant peach potato aphids in glasshouses in southern England stimulated research into biological control of these pests. At least initially, attempts to use aphid parasites were unsuccessful. However, it was realized that, if broad spectrum insecticides were to be avoided, another control for red spider mite would have to be found. Such a control was already being developed in Holland, employing a predatory mite (*Phytoseiulus* spp.) from South America. When the carbamate insecticide pirimor became available to control the aphids selectively, an integrated control programme successfully combining biological control of red spider mite with insecticidal control of the aphid was introduced.

7.3   The integrated control concept

Many other examples could be cited, particularly the successful resurrection of biological control in apple orchards of Nova Scotia over a 12-year period. Broad spectrum pesticides were generally reduced and replaced by a non-persistent plant extract insecticide (rhyania), which allowed the egg parasites of moths (the major pests) to survive. Fungicides were largely replaced by glyodin, which has little effect on the arthropod fauna.

These examples all reflect the original aim of integrated control, defined by STERN et al. (1959) as 'Applied pest control which combines and integrates biological and chemical control. Chemical control is used as necessary and in a manner which is least disruptive to biological control.' The latter sentence suggests that the chemical should be selective between the various life forms which might encounter it in the field. The ideal is obviously to use chemicals which are inherently selective, but there are good reasons for looking at ways of *making* a broader spectrum insecticide selective by the way we use it. There are some selective insecticides; for example, pirimor is highly selective for aphids. Whether used as a systemic or applied directly to insects (topical application), it shows a marked selectivity for aphids in contrast to their natural enemies when compared with some other systemic aphicides (Fig. 7–1). By and large, however, few inherently selective compounds are likely to be developed. The economic problems of developing insecticides were discussed in section 1.3.3; very few single crops or pest problems would have a large enough usage potential to warrant the development of a specific insecticide tailored to be selective for a particular integrated control problem. The second point which needs making is that too much emphasis has in the past been placed on selectivity between a pest and its natural enemies. There are many pest problems (e.g. low density pests such as disease vectors) where the pest

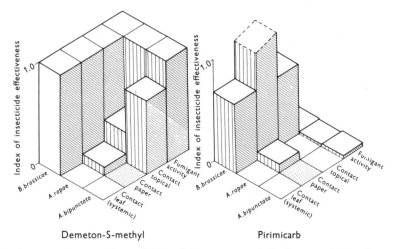

Demeton-S-methyl   Pirimicarb

Fig. 7–1   Comparative toxicities of the contact and fumigant activities of two systemic insecticides to an aphid (*Brevicoryne brassicae*), its main parasite (*Aphidius* (*Diaeretiella*) *rapae*) and a predator (the ladybird *Adalia bipunctata*). Toxicities are expressed as fractions of the effectiveness of the systemic kill of *B. brassicae* by Demeton-s-methyl. (Courtesy DODD, G. D.)

virtually has to be eliminated. This can usually only be achieved with a pesticide, and the natural enemies specific to the pest might almost just as well be killed by the pesticide as allowed to die of starvation or emigrate because of the disappearance of their prey. In such cases, integrated control involves pesticide selectivity between the pest in question and the natural enemies of other potential pests of the same crop, so that insecticide control of the key pest does not lead to the upsurge of other pest problems.

7.4   The procedure of integrated control

7.4.1.   Defining the ecosystem

Integrated control is often possible even on a small local scale, particularly in Europe where hedgerows provide a reservoir of fauna usually untreated by pesticide. However, it is most likely to be effective over a region representing a local faunal population between the patches of which intermixing occurs.

7.4.2   *Establishing economic thresholds*

This is the important decision of how far a particular pest population can be allowed to grow before insecticide must be applied to control crop loss. Insecticide applications can then be cut down to selective treatments only when absolutely necessary. STERN *et al.* (1959) defined 'economic threshold' as 'the density at which control measures should be determined to prevent an increasing pest population from reaching the economic injury level'. It is therefore a threshold for action related by experience and/or experimentation to the 'economic injury level' which is 'the lowest population level that will cause economic damage'.

The generalized relation of the 'economic injury level' to pest infestation is shown in Fig. 7–2. Low pest infestations are frequently beneficial and may stimulate plant growth or allow fewer fruits to

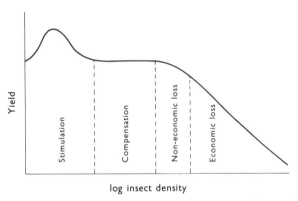

Fig. 7–2   Generalized relationship between crop yield and pest density.

develop greater size (thus alleviating the need for chemical fruit thinners). The effect of larger populations is frequently compensated for by the plant (e.g. many of the bolls produced by cotton plants will be shed in the absence of any attack; adjacent cereal plants will take up the space of killed-out plants and the larger surviving wheat plants then individually show an increased yield). Once pest damage exceeds the compensating powers of the crop, increasing populations result in a progressive reduction in yield. This damage threshold is, however, still below the economic threshold for it would not be economic to control. The costs of pesticide, labour and machinery costs as well as the damage that pesticide application normally causes (machinery damage to the crop as well as a measurable check to plant growth resulting from the toxin applied) exceed the damage caused by the pest (therefore regarded as 'non-economic damage') up to what is truly the 'economic injury level'. Beyond this the 'cost/potential benefit ratio' falls below 1. Potential benefit is, of course, not a fixed value for a given crop, because it fluctuates with the 'environmental' attitude of the grower concerned, his personal as well as local economic conditions, the state of the market for the crop, the cost of distributing the crop, the investment the crop represents and many other considerations. In spite of those fluctuations, the 'economic threshold' can usually be established sufficiently accurately to guide the decision 'to spray or not to spray' once the relationship of yield to pest density has been defined for the crop.

7.4.3 Sampling and predicting populations

The crop ecosystem must be sampled to determine whether natural mortality agents are present in sufficient numbers to be worth integrating with selective chemical control, and how frequently the economic threshold is being exceeded.

7.4.4. Augmenting the resistance of the environment

The purpose of augmentation may be to provide natural enemy action where insufficient already exists (e.g. heavily sprayed ecosystems, crops with an introduced pest) or to establish an equilibrium pest population at an artificially low level (Fig. 7–3). Introduced pests can

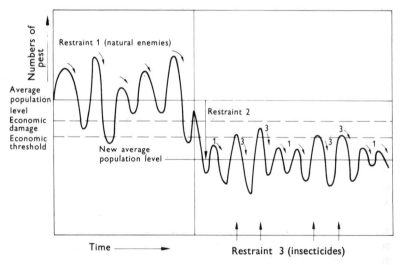

Fig. 7–3 Diagrammatic representation of the role of three restraints in an integrated control programme. Restraint 2 may be an imported natural enemy or plant resistance. (From VAN EMDEN, H. F., 1969, *J. Sci. Fd Agric.,* **20**, 385–7; courtesy of journal.)

sometimes have their populations equilibrated at a lower level by the importation of natural enemies (especially parasites) which may then reproduce the stabilizing influences of the country of origin and which the pest has escaped. Where parasites have disappeared because a vital alternative host has been lost through monoculture, replacing a single plant species (e.g. blackberries planted near vineyards in California to maintain over winter alternative prey for an effective parasite of the grape leafhopper) may be all that is necessary. If the unacceptable equilibrium is already the 'natural' one under the influence of natural enemies, adding further natural enemies is unlikely to improve the position, though it may appear to do so in the short term. Only plant resistance can have the desired effect, for the equilibrium population is related to the energy flow through the pest population and this is where plant resistance has a modifying influence.

7.4.5   Developing selective pesticide applications

The biological control potential established or already present in the environment then needs protection from the sprays necessary whenever pest populations reach the economic threshold. As has been mentioned earlier, the way we apply pesticide is likely to prove far more important than its inherent selectivity. We may be able to 'induce' selectivity by lowering the dose (cf. section 7.2.2) but even so we are seeking 'relative' rather than 'absolute' selectivity. Our application will almost certainly kill both some pests and some natural enemies; it is the balance of kill which is important to preserve a reservoir of natural enemies. Differentiating the kill of pests and natural enemies in time and space seems to offer the best possibilities.

Selectivity in space means treating only a part of the crop to enable natural enemies to survive in the untreated part. By strip-cutting alfalfa, the Californian workers were able to keep natural enemies alive in more recently cut strips whenever the adjacent taller strips required insecticide treatment. Scale insects in citrus have been controlled by allowing biological and chemical control to alternate on adjacent rows on a 2-year schedule, i.e. alternate rows were sprayed with oil emulsion and left clear in alternate years. Insecticide treated attractive baits may be used to separate the pest from the natural enemies and 'lure it to its doom' (see sections 4.3.2, 4.4).

To obtain selectivity in time requires research on the life history of the natural enemies. There may be times when a high proportion of their population is protected from contact with sprays by being inside a protective casing (e.g. in an insect egg or an aphid parasite cocooned inside a mummified aphid) or outside the treated area (e.g. in an alternative host in the hedgerow or outside the crop seeking flowers for adult feeding).

7.5   Extended concept of integrated control

The original concept of an integration of chemical and biological control has been extended by many people to embrace all suitable pest control methods integrated in a compatible manner. The widened concept of integrated control has the merit that cultural methods have been more closely examined for their control potential apart from their effect on natural enemy abundance. The danger has, however, been introduced of 'trying to avoid chemical control' rather than stressing the 'better use of chemicals' on which the original concept of integrated control depended.

7.6   Pest management

This is the current 'battle-cry' of ecologists involved in pest control and is 'the reduction of pest problems by actions selected after the life systems of the pests are understood and the ecological as well as economic consequences of these actions have been predicted, as accurately as possible, to be in the best interest of mankind' (RABB, 1970).

Pest management is therefore a 'blanket term' for an ecological approach to pest control with economic and environmental considerations very much in mind and really is a definition of what 'pest control' should and might be, rather than of a particular kind of pest control. It equally embraces the multiple approach of integrated control and single component biological control in as much as either may prove the best solution to a particular pest problem. It might be a fair statement that integrated control is likely to prove the most generally applicable pest management solution. Indeed, the principal features of pest management listed by RABB (1970) and summarized below are in several places very similar to the features of integrated control already mentioned:

(a) the *orientation* is towards entire pest populations rather than localized ones;
(b) the *proximate objective* is to lower the mean level of abundance of the pest so that fluctuations above the economic threshold are reduced or eliminated;
(c) the *method* or *combination of methods* are chosen to supplement natural control and to give the maximum long-term reliability with the cheapest and least objectionable protection;
(d) the *significance* is that alleviation of the problem is general and long-term with minimum harmful side effects;
(e) the *philosophy* is to 'manage' the pest population rather than to eliminate it.

Most effort towards the development of pest management systems has gone into investigating the contribution that might come from the modern ecological tools of life-table studies, systems analysis and mathematical modelling. If the role of the various factors which cause changes in insect abundance can be understood and related to predictable events, then a model of the system enables predictions to be made on the consequences of any agricultural or pest control practices or combinations thereof. Ideally, the system could indeed then be 'managed' to best advantage.

One obvious problem is that extensive life table data needs collecting over several years before a single pest population, let alone that of all the potentially important pests on the crop, can be modelled. Moreover, the important influence of weather resists modelling in temperate situations, yet is usually an essential ingredient of the model.

Perhaps too much stress has been laid on computer models in current pest management research. The important aspect of a model is its predictive value. Although computer models greatly speed up the production of predictions about populations, the same end can sometimes be achieved by a conceptual model based on field observation and experiment. The important step towards the 'pest management' goal is not the development of the model in itself, but the early trial of its predictions in the field, so that the pest management 'output' can be refined and improved in the light of experience in the 'real world'.

Bibliography

VAN DEN BOSCH, R. and MESSENGER, P. S. (1973). *Biological Control.* Intext Educational Publishers, New York and London

BURGES, H. D. and HUSSEY, N. W. (1971). *Microbial Control of Insects and Mites.* Academic Press, London

CARSON, R. (1962). *Silent Spring.* Houghton Mifflin Co., Boston

DOUTT, R. L. (1958). *Bull. ent. Soc. Am.,* **4,** 119–23

VAN EMDEN, H. F. (1965). *Sci. Hort.,* **17,** 121–36

HUFFAKER, C. B., ed., (1971). *Biological Control.* Plenum, New York

KNIPLING, E. F. (1955). *J. econ. Ent.,* **48,** 459–62

LEWIS, T. (1965a). *Sci. Hort.,* **17,** 74–84

LEWIS, T. (1965b). *Ann. appl. Biol.,* **55,** 513–18

METCALF, R. L. and FLINT, W. P. (1962). *Destructive and Useful Insects.* McGraw-Hill, New York and Maidenhead

PAINTER, R. H. (1951). *Insect Resistance in Crop Plants.* Macmillan, New York

RABB, R. L. (1970). Introduction to the Conference. In *Concepts of Pest Management.* Edited by R. L. Rabb and F. E. Guthrie. North Carolina State University, Raleigh

RIPPER, W. E. (1956). *A. Rev. Ent.,* **1,** 403–38

SOLOMON, M. E. (1969). *Population Dynamics.* Edward Arnold, London

SOUTHWOOD, T. R. E. (1973). The insect/plant relationship—an evolutionary perspective. In *Insect/Plant Relationships.* Edited by H. F. van Emden. Blackwell Scientific Publications, Oxford

STERN, V. M., SMITH, R. F., VAN DEN BOSCH, R. and HAGEN, K. S. (1959). *Hilgardia,* **29,** 81–101

WOLCOTT, G. N. (1942). *Science,* **96,** 317–18